Field Guide to the Canadian Forest Fire Behavior Prediction (FBP) System

3rd Edition

T0287274

S.W. Taylor
and
M.E. Alexander

SPECIAL REPORT 11

Natural Resources Canada
Canadian Forest Service
Northern Forestry Centre

2018

Catalogue No. Fo29-34/11-2018E
ISBN 978-0-660-27486-7
ISSN 1188-7419
First edition 1997
Second edition 2016
Third edition 2018

Distributed by
UBC Press
c/o UTP Distribution
5201 Dufferin Street
Toronto, ON M3H 5T8
Phone: 1-800-565-9253
Ubcpress.ca

Published by Natural Resources Canada,
Canadian Forest Service, Northern Forestry Centre,
5320–122 Street
Edmonton, AB T6H 3S5

Cette publication est également disponible en français sous le titre
*Guide de la méthode canadienne de prévision du comportement des
incendies de forêt (PCI)*

Printed in Canada

Taylor, S.W.; Alexander, M.E. 2018. Field guide to the Canadian Forest Fire
Behavior Prediction (FBP) System. 3rd Ed. Nat. Resour. Can., Can.
For. Serv., North. For. Cent., Edmonton, AB. Spec. Rep. 11.

Abstract

The Canadian Forest Fire Behavior Prediction (FBP) System is a systematic
method for assessing wildland fire behavior potential. This field guide
provides a simplified version of the system, presented in tabular format.
It was prepared to assist field staff in making first approximations of FBP
System outputs when computer-based applications are not available.
Quantitative estimates of head fire spread rate, fire intensity, type of fire,
and spread distance, elliptical fire area, perimeter, and perimeter growth
rate are provided for eighteen fuel types within five broad groupings
(coniferous, deciduous, and mixedwood forests, logging slash, and grass),
covering most of the major wildland fuel types found in Canada. The
FBP System is intended to supplement, not replace, the experience and
judgment of fire personnel.

Résumé

La Méthode canadienne de prévision du comportement des incendies
de forêt (PCI) est une méthode systématique permettant d'évaluer le
comportement potentiel des feux de forêt. Ce guide terrain fournit une
version simplifiée de la méthode en la présentant sous forme de tableaux.
Il a pour objet d'aider le personnel de terrain à établir les premières
approximations tirées de la Méthode PCI lorsqu'on ne dispose pas de
version informatisée. Les estimations quantitatives du taux de propagation
de la tête de l'incendie, de l'intensité du feu, du type d'incendie ainsi
que de la distance de propagation et du taux de croissance du périmètre
du feu sont données pour dix-huit types de combustibles correspondant
à cinq grandes catégories (conifères, feuillus, forêts mixtes, déchets de
coupe, herbe) comprenant la plupart des principaux types de combustibles
forestiers trouvés au Canada. La Méthode PCI vise à compléter, et non à
remplacer, l'expérience et le jugement des gestionnaires d'incendies.

Preface to second edition

The first version of this guide was prepared in 1995 following a number of fire behavior-related safety incidents that occurred in the 1994 fire season. Our intent then, as now, was to provide wildland firefighters with a simple means of estimating fire behavior to enhance their situational awareness on the fireline when computer aids are not readily available. The first edition has been widely used by fire crews and specialists in the field and in training courses; the ROS/fire intensity tables have been particularly useful in helping users visualize "where they are" in relation to the range of potential fire behavior and thresholds of concern.

Electronic fire behavior applications have an increasingly important place on the fire line; they will continue to become more accessible, portable, and better linked to fire weather forecasts. We hope this guide will complement those tools, continue to be a useful reference for understanding fire behavior prediction and the FBP system, and provide a simple visual aid for fire behavior situational awareness.

SWT and MEA

Acknowledgments

The Canadian Forest Fire Behavior Prediction System was developed over a 25-year period (1967–1992) by the Canadian Forest Service Fire Danger Group. We particularly acknowledge the contributions (in alphabetical order) of A.D. Kiil, B.D. Lawson, T.J. Lynham, R.S. McAlpine, S.J. Muraro, D. Quintilio, B.J. Stocks, C.E. Van Wagner, J.D. Walker, and B.M. Wotton – their dedication and commitment, along with a large number of support staff, made this guide possible.

The first edition of this guide was compiled at the suggestion of BC Forest Service Protection Branch staff, and with partial funding from the Canada-BC Partnership Agreement on Forest Resources Development. R. Pike assisted with preparing the first edition. We express our appreciation for the many helpful suggestions and comments received from fire fighters, fire behavior specialists, and instructors on the first edition.

The assistance of R. Benton and K. Hogg with the preparation of this edition is acknowledged. B. Droog, D. Finn, M. Heathcott, D. Hicks, R. Lannoville, and B. Simpson provided helpful reviews.

Disclaimer

Quick Guide to Contents

Introduction 1
Fire behavior cautions 3
Fire behavior prediction procedures 5
Fire mapping procedures 11
FBP System fuel types 14
FFMC daily drying factor 15
FFMC diurnal adjustment 21
FFMC slope aspect adjustment 23
Slope equivalent wind speed 24
Beaufort scale for estimating 10-m open wind speeds..25
Initial Spread Index/BISI 27
DMC and DC daily drying factors 28
BUI Buildup Index 29

Equilibrium rate of spread and fire intensity class
Coniferous   C-1 Spruce–lichen woodland 31
                C-2 Boreal spruce..32
                C-3 Mature jack or lodgepole pine 33
                C-4 Immature jack or lodgepole pine 34
                C-5 Red and white pine 35
                C-6 Conifer plantation..36
                C-7 Ponderosa pine/Douglas-fir 38
Deciduous   D-1 Leafless aspen 39
                D-2 Green aspen 40
Mixedwood M-1 Boreal mixedwood – leafless 41
                M-2 Boreal mixedwood – green..44
                M-3 Dead balsam fir mixedwood – leafless......................................47
                M-4 Dead balsam fir mixedwood – green..52
Open   O-1a Matted grass 53
                O-1b Standing grass 54
Slash   S-1 Jack or lodgepole pine slash 55
                S-2 White spruce/balsam slash 56
                S-3 Coastal cedar/hemlock/Douglas-fir slash 57

Spread distance..58
Fire area and perimeter 60
Perimeter growth rate..62
Flank fire spread rate..64
References 65
Appendix 1. Abbreviations..68
Appendix 2. Glossary..69
Appendix 3. Selected unit conversion factors..77
Appendix 4. FBP System fuel type photographs 78
Appendix 5. Summary of FBP System fuel type characteristics..95
Appendix 6. Flame length fire intensity class for surface fires..98
Appendix 7. Fire behavior descriptions..99
Appendix 8. Fire behavior prediction worksheet 100

Contents

Introduction.. 1

    Assumptions... 2

    Accuracy of the guide... 3

    Fire behavior cautions 3

Fire behavior prediction procedure 5

Fire mapping procedures.. 11

FBP System fuel types... 14

FFMC daily drying tables 15

FFMC diurnal adjustment.. 21

FFMC slope aspect adjustment... 23

Slope equivalent wind speed.. 24

Beaufort scale for estimating 10-m open wind speed........... 25

Wind speed adjustment 26

Initial Spread Index/BISI 27

DMC and DC daily drying factors 28

Buildup Index.. 29

Equilibrium rate of spread and fire intensity class................. 31

Spread distance 58

Fire area and perimeter 60

Perimeter growth rate... 62

Flank fire spread rate 64

References... 65

Tables

1.   Conditions to watch for and related fire behavior interpretations 4

2.   FBP System fuel types 14

3.1.   Daily Fine Fuel Moisture Codes for 10.5°–15°C 15

3.2.   Daily Fine Fuel Moisture Codes for 15.5°–20°C 16

3.3.	Daily Fine Fuel Moisture Codes for 20.5°–25°C	17
3.4.	Daily Fine Fuel Moisture Codes for 25.5°–30°C	18
3.5.	Daily Fine Fuel Moisture Codes for 30.5°–35.5°C	19
3.6.	Daily Fine Fuel Moisture Codes for >35.5°C	20
4.1.	Diurnal Fine Fuel Moisture Codes for afternoon and overnight	21
4.2.	Diurnal Fine Fuel Moisture Codes for morning	22
4.3.	Fine Fuel Moisture Code slope aspect adjustment	23
5.	Slope equivalent wind speed	24
6.1.	Beaufort scale for estimating 10-m open wind speed	25
6.2.	Wind speed adjustment factor to 10-m height	26
6.3.	Initial Spread Index and Backfire Initial Spread Index	27
7.1.	Duff Moisture Code daily drying factors	28
7.2.	Drought Code daily drying factors	28
8.1.	Buildup Index for Drought Codes of 0–224	29
8.2.	Buildup Index for Drought Codes of 225–599	30
9.1.	Equilibrium rate of spread and fire intensity class for C-1 spruce–lichen woodland	31
9.2.	Equilibrium rate of spread and fire intensity class for C-2 boreal spruce	32
9.3.	Equilibrium rate of spread and fire intensity class for C-3 mature jack or lodgepole pine	33
9.4.	Equilibrium rate of spread and fire intensity class for C-4 immature jack or lodgepole pine	34
9.5.	Equilibrium rate of spread and fire intensity class for C-5 red and white pine	35
9.6.	Equilibrium rate of spread and fire intensity class for C-6 conifer plantation with 7-m crown base height	36
9.7.	Equilibrium rate of spread and fire intensity class for C-6 conifer plantation with 2-m crown base height	37
9.8.	Equilibrium rate of spread and fire intensity class for C-7 ponderosa pine/Douglas-fir	38

9.9.   Equilibrium rate of spread and fire intensity class for
        D-1 leafless aspen 39

9.10.   Equilibrium rate of spread and fire intensity class for
        D-2 green aspen 40

9.11.   Equilibrium rate of spread and fire intensity class for
        M-1 boreal mixedwood—leafless (75% conifer, 25% deciduous) 41

9.12.   Equilibrium rate of spread and fire intensity class for
        M-1 boreal mixedwood—leafless (50% conifer, 50% deciduous) 42

9.13.   Equilibrium rate of spread and fire intensity class for
        M-1 boreal mixedwood—leafless (25% conifer, 75% deciduous) 43

9.14.   Equilibrium rate of spread and fire intensity class for
        M-2 boreal mixedwood—green (75% conifer, 25% deciduous) 44

9.15.   Equilibrium rate of spread and fire intensity class for
        M-2 boreal mixedwood—green (50% conifer, 50% deciduous) 45

9.16.   Equilibrium rate of spread and fire intensity class for
        M-2 boreal mixedwood— green (25% conifer, 75% deciduous) 46

9.17.   Equilibrium rate of spread and fire intensity class for
        M-3 dead balsam fir mixedwood—leafless (30% dead fir) 47

9.18.   Equilibrium rate of spread and fire intensity class for
        M-3 dead balsam fir mixedwood—leafless (60% dead fir) 48

9.19.   Equilibrium rate of spread and fire intensity class for
        M-3 dead balsam fir mixedwood—leafless (100% dead fir) 49

9.20.   Equilibrium rate of spread and fire intensity class for
        M-4 dead balsam fir mixedwood—green (30% dead fir) 50

9.21.   Equilibrium rate of spread and fire intensity class for
        M-4 dead balsam fir mixedwood—green (60% dead fir) 51

9.22.   Equilibrium rate of spread and fire intensity class for
        M-4 dead balsam fir mixedwood—green (100% dead fir) 52

9.23   Equilibrium rate of spread and fire intensity class for
        O-1a matted grass 53

9.24   Equilibrium rate of spread and fire intensity class for
        O-1b standing grass 54

9.25.   Equilibrium rate of spread and fire intensity class for
        S-1 jack or lodgepole pine slash 55

9.26.	Equilibrium rate of spread and fire intensity class for S-2 white spruce/balsam slash	56
9.27.	Equilibrium rate of spread and fire intensity class for S-3 coastal cedar/hemlock/Douglas-fir slash	57
10.1.	Spread distance for all fuel types with equilibrium spread rate, and for open fuel types and surface fires in closed fuel types with accelerating rate of spread	58
10.2.	Spread distance for crown fires in closed fuel types with accelerating rate of spread	59
11.1	Fire area and perimeter for conifer, deciduous, mixedwood, and slash fuel types	60
11.2.	Fire area and perimeter for O-1 matted and standing grass	61
12.1.	Perimeter growth rate for conifer, deciduous, mixedwood, and slash fuel types	62
12.2.	Perimeter growth rate for grass fuel types	63
13.	Flank fire spread rate	64

Figures

1.	Flow chart for procedures used in this guide.	7
2.	Vector addition procedure for estimating effective wind speed and spread azimuth when the wind is blowing.	8
3.	Simplified fire growth scenarios.	13

Appendixes

1. Abbreviations	68
2. Glossary	69
3. Selected unit conversion factors	77
4. FBP System fuel type photographs	78

Coniferous

C-1 Spruce–lichen woodland	78
C-2 Boreal spruce	79
C-3 Mature jack or lodgepole pine	80

C-4 Immature jack or lodgepole pine... 81

C-5 Red and white pine.. 82

C-6 Conifer plantation 83

C-7 Ponderosa pine/Douglas-fir 84

Deciduous

D-1 Leafless aspen... 85

D-2 Green aspen 86

Mixedwood

M-1 Boreal mixedwood—leafless 87

M-2 Boreal mixedwood—green... 88

M-3 Dead balsam fir/mixedwood—leafless.. 89

M-4 Dead balsam fir/mixedwood—green 90

Open

O-1 Grass... 91

Slash

S-1 Jack or lodgepole pine slash... 92

S-2 White spruce/balsam slash 93

S-3 Coastal cedar/hemlock/Douglas-fir slash 94

5. Summary of FBP System fuel type characteristics................................... 95

6. Flame length and fire intensity class for surface fires 98

7. Fire behavior descriptions... 99

8. Fire behavior prediction worksheet example.. 100

    Fire behavior prediction worksheet.. 101

x

Introduction

The Canadian Forest Fire Behavior Prediction (FBP) System provides a systematic way to estimate the potential behavior of wildland fires. It is made up of a set of mathematical equations relating fire behavior characteristics to wind, fuel moisture, and topographic conditions for 18 fuel (vegetation) types in Canada (Forestry Canada Fire Danger Group 1992; Wotton et al. 2009) and further interpretations for leafed aspen (Alexander 2010).

Complete and precise FBP System predictions are best made using computer-based applications. However, this guide may be used by persons with training in fire behavior and the FBP System as a reference when electronic applications are unavailable. It is also helpful in visualizing changes in fire intensity class.

The core of the FBP System field guide is the rate of spread (ROS) and fire intensity class tables, which are essentially inverted "hauling charts" (Andrews and Rothermel 1982). They allow the user to determine a fire's ROS, intensity level, and general type of fire (i.e., surface, intermittent crown, or continuous crown fire) in one simple look-up procedure. The charts provide a visual representation of how the current conditions relate to the range of potential fire behavior. This is particularly important in conifer forests where the transition of a fire from surface to canopy occurs over a narrow range of conditions.

Further guidance regarding the use of the FBP System can be found in a number of guides (e.g., Hirsch 1996; Pearce et al. 2008; Kidnie et al. 2010), interactive training materials, and training courses. Fire suppression interpretations based on head fire intensity class are also available for several fuel types (Alexander and De Groot 1998; Alexander and Lanoville 1989; Cole and Alexander 1995).

While the FBP System has proven to be useful in practical fire management decision making, users should understand that no matter how good a model is, it is often not possible to predict fire behavior with a high degree of precision. This is because it is difficult, when using models, to measure or represent variable fuel conditions (including fuel moisture content) across a landscape at high resolution; furthermore, it is not possible to know the state of the atmosphere and predict with great accuracy future weather conditions such as wind speed across a landscape and over time. It is more useful to consider the range of potential fire behavior over a forecast period.

Assumptions

Users must be careful not to apply the system beyond its useful range. The FBP System can be used to make predictions for a fire spreading during one burning period from a single point source or line of fire, with the following assumptions:

- Fuel conditions are similar to one of the 18 benchmark fuel types.

- The fuel moisture codes used are representative of the site conditions.

- Fuels are uniform and continuous, topography is simple and homogenous, and the wind is constant and unidirectional during the burning period.

- The fire is wind or wind/slope driven, and spread is not unduly affected by a convection column. Wind is measured in the open, at or corrected to 10 m.

- The rate of fire spread levels off at very high wind speed and initial spread index (ISI) values.

- The fire is unaffected by suppression activities (free burning).

- A fire starting from a point source will have an elliptical shape under the above conditions.

- The effect of short-range spotting of firebrands on spread is taken into account.

The FBP System is based on observations of experimental fires and wildfires. There are very few records of sustained fire spread with wind speeds above 60 km and/or ISI > 70. Thus the largest ROS values in this guide roughly correspond to the upper limit of observed spread rates in each fuel type. Higher ROS values (>200 m/min in conifers and > 350 m/ min in grass) may occur with higher wind speeds and during gusts.

Additionally, predictions may be made for more complex fuel and topographic situations, or for changing fuel moisture or wind conditions over time, by breaking the landscape or burning period into discrete portions.

Accuracy of the guide

While every effort was made to ensure that this guide represents the FBP System as accurately as possible, some simplifications were made in order to present important fire behavior characteristics in table form. The guide will give a good estimate of Fire Intensity Class and Type of Fire in most cases. Users should be able to estimate fire size characteristics within ±20% of computed values for rate of spread >3 m/min in most fuel types. Fire size estimates are less accurate for ROS <3 m/min because of interpolation and rounding errors; however, the predictions are still of practical value.

Fire behavior cautions

Fire behavior prediction systems are intended to assist firefighters and officers in estimating potential fire behavior in constant conditions—they complement but do not replace training, experience, sound judgment, and observations of on-going fire behavior. No system can ever fully account for all of the variables that affect fire behavior. The FBP System does not, for example, represent the effects of the following factors on fire behavior: seasonal changes in the moisture content of live understory vegetation, atmospheric stability/instability, interactions between the convection column and the atmosphere, or long-range spotting.

Fire behavior predictions are usually made for a particular set of conditions and may not signal changes and transitions in fire behavior associated with rapidly changing weather (see Lawson and Armitage 2008), topography, and fuel conditions. Users must be aware of the limitations of the system, anticipate transitions, and watch for unusual situations. Table 1 lists some important situations to watch for.

Table 1. Conditions to watch for and related fire behavior interpretations

Conditions		Fire behavior interpretations
Weather	90° shift in wind direction	Flank of fire becomes head of fire.
	Approaching dry cold front	Potential for low-level jet at height of convection column, increasing convective circulation and surface spread.
	Approaching thunderstorm	Potential for downdrafts with increasing wind speed and spread rate.
	Humidity < temperature (crossover)	Dry, light fuels, increased ignition potential, and spread rate.
	Combined effects of increasing temperature and wind speed with decreasing relative humidity	Potential for rapid spread.
	Slope and valley winds	Increased spread and intensity upslope/up valley with daytime heating, downslope/down valley after dark.
Fuels	Light, open fuels	Potential for rapid initial spread.
	Dead trees or tops	Long-range spotting, falling trees.
	Fuel type changes	Change in spread rate and intensity
Topography	Complex topography (gullies, canyons, ridge tops)	Turbulent winds (funneling in gullies; lee waves and eddies at ridges) and fire spread.
	Steep slopes > 50%	Flame attachment and rapid spread; burning debris rolling downslope.
Fire Behavior	Smoke and haze layer	Greater spread and intensity above than below the inversion layer.
	Increase in smoke venting	Atmospheric inversion lifting: increase in spread, intensity.
	Dust devils and fire whirls	Surface level instability, erratic spread.
	Black smoke above surface fire	Torching, potential transition to crown fire.
	Towering convection column	Intense burning, fire whirls, long-range spotting.
	Long range spotting	Strong updrafts in convection column. Spread rate difficult to predict. New fires across fire guards and topographic barriers.

4

Fire behavior prediction procedure

A procedure to determine several key fire characteristics using this guide is outlined below. Figure 1 provides a flow chart for procedures used in this guide. Repeat the procedure (separate prediction points) for each different fuel type or slope/aspect class that may be encountered during the prediction time interval. The abbreviations used are listed in Appendix 1 and many of the terms are defined in the Glossary (Appendix 2). Some correction and conversion factors are given in Appendix 3. Appendix 4 provides FBP fuel type photographs. Appendix 5 comprises a summary of FBP System fuel type characteristics. Appendix 6 relates flame dimensions to fire intensity. Appendix 7 provides fire behavior descriptions. The fire behavior prediction worksheet in the back of the guide may be used to record the input data, intermediate computations, and the resulting fire characteristics. A filled-out example of the worksheet is also provided in Appendix 8.

Line no(s).

1–3	Enter the **fire number or name**; **date** and **time**; **prediction date** and **interval**; and the **ignition type** for each prediction point: point ignition (PI) for initiating fires or line source (LS) for fires spreading from an established perimeter.
4	Select the most appropriate **fuel type** (Table 2) and enter the fuel type **identifier** that best represents the fuels at the prediction point. In choosing a fuel type, the physical properties of the fuel complex such as surface fuel conditions, stand density, and crown base height should be considered in addition to the tree species composition.
5	Record any relevant fuel type **modifiers**: for C-6 enter **Crown Base Height (CBH)**; for M-1 and M-2 enter the Percent Conifer (PC) / Percent Deciduous (PD) composition (e.g., 75:25); and for M-3 and M-4 enter Percent Dead Fir (PDF).
6	Enter the **standard daily Fine Fuel Moisture Code (FFMC)** for the prediction date.

Optional: Tables are provided to calculate the daily FFMC for days without rain. Find the appropriate table for the expected temperature in Tables 3.1–3.6. Find yesterday's FFMC in the top row, and move down the column to the expected RH and wind speed range row and get today's FFMC.

7   Enter the **hourly FFMC** (which is calculated from hourly weather observations) for the prediction date/time, if available. Alternatively, determine the **diurnal FFMC** for the prediction time from the standard daily FFMC (Tables 4.1 or 4.2).

8–10   Enter the **ground slope** (%) and **aspect** (cardinal direction) for the area ahead of the prediction point, disregarding minor variations. For slash and grass fuel types, also record the slope and aspect of the weather observation point in brackets (use L for level): e.g., Ground slope 50 [15]; Aspect S [E]. These values may be used to determine a topographically **adjusted FFMC** for open fuel types on clear days during the months of March, April, August, September, or October between 1200 and 2000 LST (Table 4.3).

11   Determine the **slope equivalent wind speed** corresponding to the ground slope for the selected fuel type (Table 5).

12–13   Enter the 10-m open wind speed as measured, estimated, or forecast for the prediction point. Table 6.1 (Beaufort Scale) may be used as a guide if instruments are not available. Use Table 6.2 to correct for measurements at non-standard heights. Enter the wind speed as a positive value if the wind is blowing upslope (e.g., 23 km/h) or a negative value if it is blowing downslope (e.g., -3 km/h). Determine the effective wind speed (i.e., 10-m open wind + slope equivalent wind). If the wind is blowing downslope the result may be positive or negative. Note that if the effective wind speed is negative, the fire is predicted to spread downslope.

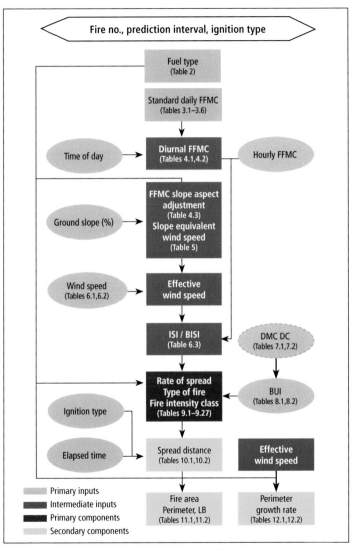

Figure 1. Flow chart of procedures used in this guide.

Optional: The interaction between wind and slope effects on the effective wind speed and the spread azimuth can be estimated graphically using vector addition (Fig. 2) when the wind is blowing A. upslope B. downslope or C. across slope. In each case,

i) plot the wind speed vector with the distance scaled to wind speed at the appropriate wind azimuth.

ii) starting at the head of the wind speed vector, plot the slope equivalent wind speed vector at the same scale using the appropriate slope azimuth (opposite of the slope aspect).

iii) starting at the tail of the wind speed vector, plot the line completing the triangle. This is the effective wind speed vector. Measure the vector length and convert to wind speed (km/h) with the appropriate scale conversion, and measure the spread azimuth. In Figure 2 C i) wind speed is 10 km/h; wind azimuth is 80° ii) slope equivalent wind speed is 4 km/h; slope azimuth is 360°, iii) effective wind speed is 11.2 km/h; spread azimuth is 62°.

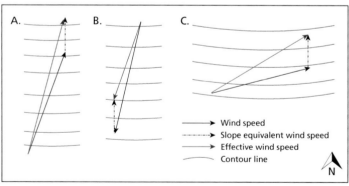

Figure 2. Vector addition procedure for estimating effective wind speed and spread azimuth when the wind is blowing A) upslope, B) downslope, and C) across slope.

14   Find the head and backfire **Initial Spread Index (ISI)** from the FFMC and effective wind speed (Table 6.3).

15   Enter the **Buildup Index (BUI)** for the prediction date from the closest representative fire weather station, or the **degree of curing** (%) for the O-1 fuel types.

Optional: Today's **BUI** can be calculated (for days without rain) using Tables 7.1, 7.2, 8.1 and 8.2. Find the **DMC** and **DC** drying factors for the appropriate month and the weather conditions in Table 7.1 and 7.2. Add the drying factors to yesterday's DMC and DC to find today's DMC and DC. Find today's BUI in Table 8.1 and 8.2. For rain days, use a computer application or the full set of **Fire Weather Index System** tables (CFS 1984).

16–20   For each fuel type, find the **equilibrium rate of spread (ROS)**, **fire intensity class**, **type of fire**, and **crown fraction burned CFB) level** for both the head fire and the backfire (if relevant) from the head fire and backfire ISI and the BUI or degree of curing (Tables 9.1–9.27). BUI_0 (the standard BUI assumed in the ROS model and shown as the boxed value in the BUI column heading) may be used if the BUI is not known. The background shading in the tables indicates fire intensity class. The type of fire is indicated as follows:

Black values indicate	surface fire with	<10% CFB
Black values with * indicate	intermittent crown fire with	10–89% CFB
White values indicate	continuous crown fire with	>90% CFB

Record the type of fire as **S (surface fire)**, **IC (intermittent crown fire)**, or **CC (continuous crown fire)**. The line within the intermittent crown fire class is the 50% **CFB (crown fraction burned)** level. For surface fire (S) enter CFB as < 10%, for continuous crown fire (CC) enter >90% CFB, and for intermittent crown fire select the closest CFB level: 10, 50, or 90%.

21–24   Enter the **elapsed time** corresponding to the prediction time interval. Determine the **head fire spread distance**, **backfire spread distance**, and **total spread distance** (head fire + backfire spread distances) from Table 10.1 or 10.2. Distances are given for three spread functions:

- equilibrium ROS: all fuel types

- accelerating ROS: open fuels and surface fires in closed fuel types (<10% CFB)

- accelerating ROS: crown fires in closed fuel types (50% or 90% CFB)

Use the equilibrium ROS function for fires spreading from an active fire perimeter or other line ignition (LS). Use the acceleration function for fires spreading from a point ignition type (PI). Determine the backfire spread distance only if there are no barriers to backfire spread.

In Table 10.1 Note that the 5 min column defines a higher risk "deadman" zone if the wind direction shifts and a flank fire becomes a head fire.

Note that where the elapsed times correspond to the time needed for a tactical withdrawl or evacuation (including a safety margin), the table values represent the distance to trigger points.

25–27   For fires starting from a point source, determine the **elliptical fire area**, **elliptical fire perimeter** and the **length/breadth ratio (L/B)** from the total spread distance (head + back) and the effective wind speed during the prediction interval (Table 11.1 or 11. 2). An elliptically shaped fire's maximum width or breadth can be found by dividing the total spread distance by the L/B ratio.

28   The **perimeter growth rate** can be determined from the sum of the head and backfire ROS and the effective wind speed in Table 12.1 or 12.2. Values are given for both timber/slash and grass fuel types.

29   Optional: The **flank fire spread rate** can be determined from the sum of the head and backfire ROS and the effective wind speed in Table 13.

Fire mapping procedures

The procedures below can be used to provide maps and initial estimates of fire size and perimeter for single or multiple burning periods when fire growth simulation models are not readily available[1]. Use the procedures for the most applicable scenario A–C in Figure 3.

A.   Spread from a point ignition over one burning period

1)   Convert the total (head and backfire) spread distance (m) to a map distance (cm) by dividing by the denominator of the metric map scale and multiply by 100 (e.g., 1000 m spread distance on a 1:50000 map is 1000 m / 50 000 X 100 = 2 cm).

---

[1] The elliptical fire growth model has been used mainly to estimate fire size for a single burning period (e.g., Rothermel 1991). Because the assumptions of the model are challenged with increasing fire size, computer simulation models such as Prometheus (Tymstra et al. 2010) have been developed to project fire growth over longer and multiple burning periods in complex terrain and fuel conditions. However, it is often necessary to quickly assess the potential fire growth, not only on the present day, but over several potential burning periods, and simple mapping methods can provide useful first approximations.

2) Calculate the flank fire spread distance by dividing the total spread distance by L/B.

3) Plot the head and backfire distances through the ignition point with the appropriate spread azimuth—the end points of this major semi-axis are the vertices, and the mid-point is the center of the ellipse. If there is no slope effect, the spread azimuth is the wind direction + 180° (for WD ≤ 180°) and wind direction − 180° (for WD > 180°.

4) Plot the flank fire spread distance bisecting the center; the end points of the flank fire semi-axis are the co-vertices.

5) Plot the perimeter connecting the vertices and co-vertices.

B. *Spread over multiple burning periods in a constant direction*

If the projected spread direction is the same for successive intervals, then fire spread distances are cumulative. Plot the successive head and backfire spread distances from the vertices of the previous ellipse, as show in (B). Determine the flank fire spread distance for each interval by dividing the total cumulative spread distance over all intervals by L/B; plot the flank fire spread distance from the center of the new ellipse (mid-point of the total head + backfire spread semi-axis).

The perimeter and area after each interval can be estimated by using the total cumulative spread distance in Table 11.1–11.2 and the most representative L/B. In the special case where ROS is constant in each burning period, the fire perimeter length increases linearly following the series 1, 2, 3, etc. and fire size (area) increases following the series 1:4:9:16, etc. over equal intervals (e.g., the fire perimeter will be twice as long and the area burned four times as large after the second burning period than after the first).

C. *Spread over multiple burning periods with shifting direction*

1) For the period following a wind shift plot ellipses representing the head fire spread parallel to the new spread direction in a chain-like manner at the vertices on the head and back of the previous ellipse, and the co-vertex on the leeward (down wind) flank, as shown in (C). This is a simplification of the methods used in fire growth simulation models such as Prometheus.

2) Plot the backfire spread distance from the new windward (up wind) flank.

3) Plot the new fire perimeter as the convex hull of the set of all ellipses. (Visualize an elastic band stretched around the ellipses.) Perimeter length and area have to be estimated using map measuring devices (e.g., dot grid) in the following manner:

Length (m) = map distance (cm) X map scale denominator/100 (e.g., for 1/50 000 scale, 10 cm X 50 000/100 = 5 000 m);
Area (ha) = map area (cm^2) x (map scale denominator/10 000)^2 (e.g., for 1/50 000 scale, 10 cm^2 x (50 000/10 000)^2 = 250 ha.

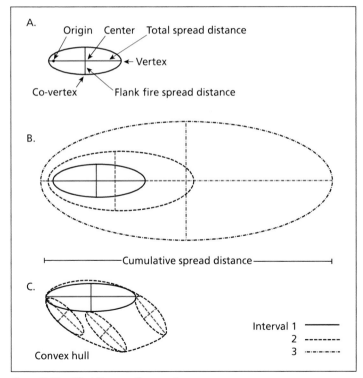

Figure 3. Simplified fire growth scenarios.

Table 2. FBP System fuel types

Group/Identifier	Descriptive name	Exposure
Coniferous (C)		
C-1	Spruce–lichen woodland	Open
C-2	Boreal spruce	Closed[a]
C-3	Mature jack or lodgepole pine	Closed
C-4	Immature jack or lodgepole pine	Closed
C-5	Red and white pine	Closed
C-6[b]	Conifer plantation	Closed
C-7	Ponderosa pine/Douglas-fir	Closed[a]
Deciduous (D)		
D-1	Leafless aspen	Open
D-2	Green aspen	Open
Mixedwood (M)		
M-1[c]	Boreal mixedwood—leafless	Closed
M-2[c]	Boreal mixedwood—green	Closed
M-3[d]	Dead balsam fir mixedwood—leafless	Closed
M-4[d]	Dead balsam fir mixedwood—green	Closed
Open		
O-1a[e]	Matted grass	Open
O-1b[e]	Standing grass	Open
Slash (S)		
S-1	Jack or lodgepole pine slash	Open
S-2	White spruce/balsam slash	Open
S-3	Coastal cedar/hemlock/Douglas-fir slash	Open

[a]C-2 and C-7 are considered open if crown closure < 50%.
[b]Crown base height can vary (see Tables 9.6 and 9.7).
[c]User must specify percent conifer composition (for M-1, see Tables 9.11, 9.12, and 9.13; for M-2, see Tables 9.14, 9.15, and 9.16).
[d]User must specify percent dead fir (for M-3, see Tables 9.17, 9.18, and 9.19; for M-4, see Tables 9.20, 9.21, and 9.22.).
[e]User must specify degree of curing and can specify fuel load.

Table 3.1.

FFMC daily

10.5–15 °C

Yesterday's FFMC

RH (%)	Wind (km/h)	80	81	82	83	84	85	86	87	88	89	90	91	92	93	94	95	96	97	98	99
0–10	0–3	91	91	92	92	93	93	93	94	94	94	95	95	95	96	96	96	97	97	98	99
	4–13	92	93	93	93	93	94	94	94	95	95	95	95	96	96	96	97	97	97	98	99
	14+	93	94	94	94	94	95	95	95	95	95	96	96	96	96	97	97	97	97	98	99
11–18	0–3	89	89	90	90	90	91	91	91	92	92	92	93	93	94	94	95	96	96	97	98
	4–13	90	90	90	91	91	91	92	92	92	92	93	93	93	94	94	95	96	96	97	97
	14+	91	91	91	91	92	92	92	92	93	93	93	93	93	94	94	95	95	96	96	97
19–28	0–3	87	88	88	88	89	89	89	90	90	91	91	91	92	93	94	94	95	95	96	96
	4–13	88	88	89	89	89	90	90	90	90	91	91	91	92	93	93	94	94	95	95	96
	14+	89	89	89	90	90	90	90	90	91	91	91	91	92	93	93	94	94	94	95	95
29–38	0–3	86	86	86	87	87	88	88	88	89	89	90	91	91	92	92	93	93	94	94	95
	4–13	86	87	87	87	88	88	88	89	89	89	90	91	91	92	92	92	93	93	94	94
	14+	87	87	88	88	88	88	89	89	89	89	90	91	91	91	92	92	92	93	93	93
39–49	0–3	84	85	85	86	86	86	87	87	88	89	89	90	90	91	91	92	92	93	93	93
	4–13	85	85	86	86	86	87	87	87	88	89	89	90	90	90	91	91	91	92	92	92
	14+	85	86	86	86	86	87	87	87	88	89	89	89	90	90	90	91	91	91	91	92
50–61	0–3	83	83	84	84	85	85	86	87	87	88	88	89	89	89	90	90	91	91	91	92
	4–13	83	84	84	84	85	85	86	87	87	88	88	88	89	89	89	89	90	90	90	91
	14+	84	84	84	85	85	85	86	87	87	87	88	88	88	88	89	89	89	89	90	90
62–73	0–3	82	82	83	83	84	85	85	86	86	86	87	87	87	88	88	89	89	89	90	90
	4–13	82	82	83	83	84	85	85	85	86	86	86	87	87	87	87	88	88	88	89	89
	14+	82	83	83	83	84	85	85	85	86	86	86	86	86	87	87	87	87	87	88	88
74–84	0–3	80	81	82	83	83	83	84	84	84	85	85	85	86	86	86	87	87	87	88	88
	4–13	80	81	82	83	83	83	83	84	84	84	84	85	85	85	85	86	86	86	86	87
	14+	81	81	82	83	83	83	83	83	84	84	84	84	84	85	85	85	85	85	86	86
85–93	0–3	80	80	80	81	81	81	82	82	82	83	83	83	83	84	84	84	85	85	85	85
	4–13	80	80	80	80	81	81	81	81	82	82	82	82	83	83	83	83	83	84	84	84
	14+	80	80	80	80	80	81	81	81	81	81	81	82	82	82	82	82	83	83	83	83
94–100	0–3	77	77	77	78	78	78	79	79	79	79	80	80	80	81	81	81	81	82	82	82
	4–13	76	77	77	77	77	78	78	78	78	79	79	79	79	79	80	80	80	80	81	81
	14+	76	76	77	77	77	77	77	77	78	78	78	78	78	78	79	79	79	79	79	79

Red values indicate fine fuels are drying.   Black values indicate fine fuels are stable or wetting.

Table 3.2.

FFMC daily

15.5–20 °C

Yesterday's FFMC

RH (%)	Wind (km/h)	80	81	82	83	84	85	86	87	88	89	90	91	92	93	94	95	96	97	98	99
0–10	0–3	92	93	93	93	94	94	94	95	95	95	95	96	96	96	97	97	97	97	98	99
	4–13	94	94	94	94	95	95	95	95	96	96	96	96	96	97	97	97	97	97	98	99
	14+	95	95	95	95	95	96	96	96	96	96	96	97	97	97	97	97	97	97	98	99
11–18	0–3	90	91	91	91	91	92	92	92	93	93	93	94	94	94	94	95	96	97	97	98
	4–13	91	92	92	92	92	92	93	93	93	93	94	94	94	94	94	95	96	96	97	97
	14+	92	92	92	93	93	93	93	93	93	94	94	94	94	94	94	95	96	96	97	97
19–28	0–3	88	89	89	89	90	90	90	91	91	91	92	92	92	93	94	94	95	95	96	97
	4–13	89	90	90	90	90	91	91	91	91	92	92	92	92	93	94	94	95	95	95	96
	14+	90	90	90	91	91	91	91	91	92	92	92	92	92	93	94	94	94	95	95	95
29–38	0–3	87	87	87	88	88	88	89	89	89	90	90	91	92	92	93	93	94	94	94	95
	4–13	88	88	88	88	89	89	89	89	90	90	90	91	92	92	92	93	93	93	94	94
	14+	88	88	89	89	89	89	89	90	90	90	90	91	92	92	92	92	93	93	93	93
39–49	0–3	85	86	86	86	87	87	87	88	88	89	90	90	91	91	91	92	92	93	93	93
	4–13	86	86	86	87	87	87	88	88	88	89	90	90	90	91	91	91	92	92	92	92
	14+	87	87	87	87	87	88	88	88	88	89	90	90	90	90	91	91	91	91	91	92
50–61	0–3	84	84	84	85	85	86	86	87	88	88	88	89	89	89	90	90	90	91	91	91
	4–13	84	85	85	85	86	86	86	87	88	88	88	89	89	89	89	90	90	90	90	90
	14+	85	85	85	85	86	86	86	87	88	88	88	88	88	89	89	89	89	89	90	90
62–73	0–3	82	83	83	84	84	85	86	86	86	87	87	87	88	88	88	88	89	89	89	90
	4–13	83	83	83	84	84	85	86	86	86	86	87	87	87	87	87	88	88	88	88	89
	14+	83	83	84	84	84	85	86	86	86	86	86	87	87	87	87	87	87	87	88	88
74–84	0–3	81	81	82	83	83	84	84	84	85	85	85	85	86	86	86	86	87	87	87	87
	4–13	81	81	82	83	83	84	84	84	84	84	85	85	85	85	85	86	86	86	86	86
	14+	81	81	82	83	83	84	84	84	84	84	84	84	85	85	85	85	85	85	85	86
85–93	0–3	80	80	81	81	81	81	82	82	82	82	83	83	83	83	84	84	84	84	85	85
	4–13	80	80	81	81	81	81	81	82	82	82	82	82	82	83	83	83	83	83	83	84
	14+	80	80	80	81	81	81	81	81	81	81	82	82	82	82	82	82	82	82	83	83
94–100	0–3	77	77	77	78	78	78	78	79	79	79	79	80	80	80	80	81	81	81	81	81
	4–13	77	77	77	77	77	78	78	78	78	78	79	79	79	79	79	79	79	80	80	80
	14+	77	77	77	77	77	77	77	77	78	78	78	78	78	78	78	78	79	79	79	79

Red values indicate fine fuels are drying.   Black values indicate fine fuels are stable or wetting.

Table 3.3.

FFMC daily

20.5–25 °C

Yesterday's FFMC

RH (%)	Wind (km/h)	80	81	82	83	84	85	86	87	88	89	90	91	92	93	94	95	96	97	98	99
0–10	0–3	94	94	94	95	95	95	95	96	96	96	96	97	97	97	97	97	98	98	98	99
	4–13	95	95	95	96	96	96	96	96	96	97	97	97	97	97	97	98	98	98	98	99
	14+	96	96	96	96	96	97	97	97	97	97	97	97	97	98	98	98	98	98	98	99
11–18	0–3	92	92	92	92	93	93	93	93	94	94	94	94	95	95	95	95	96	97	97	98
	4–13	93	93	93	93	93	94	94	94	94	94	94	95	95	95	95	95	96	97	97	98
	14+	94	94	94	94	94	94	94	94	94	95	95	95	95	95	95	95	96	97	97	97
19–28	0–3	90	90	90	91	91	91	91	92	92	92	92	93	93	93	94	95	95	96	96	97
	4–13	91	91	91	91	92	92	92	92	92	92	93	93	93	93	94	95	95	95	96	96
	14+	91	92	92	92	92	92	92	92	92	93	93	93	93	93	94	95	95	95	95	96
29–38	0–3	88	88	89	89	89	89	90	90	90	90	91	91	92	93	93	93	94	94	95	95
	4–13	89	89	89	89	90	90	90	90	90	91	91	91	92	93	93	93	93	94	94	94
	14+	90	90	90	90	90	90	90	90	91	91	91	91	92	93	93	93	93	93	93	94
39–49	0–3	86	87	87	87	88	88	88	88	89	89	90	91	91	91	92	92	92	93	93	93
	4–13	87	87	88	88	88	88	88	89	89	89	90	91	91	91	91	92	92	92	92	92
	14+	88	88	88	88	88	88	89	89	89	89	90	91	91	91	91	91	91	92	92	92
50–61	0–3	85	85	85	86	86	86	87	87	88	89	89	89	89	90	90	90	91	91	91	91
	4–13	85	86	86	86	86	86	87	87	88	89	89	89	89	89	90	90	90	90	90	90
	14+	86	86	86	86	86	87	87	87	88	89	89	89	89	89	89	89	89	90	90	90
62–73	0–3	83	83	84	84	85	85	86	87	87	87	87	88	88	88	88	88	89	89	89	89
	4–13	83	84	84	84	85	85	86	87	87	87	87	87	87	88	88	88	88	88	88	88
	14+	84	84	84	84	85	85	86	86	87	87	87	87	87	87	87	87	87	88	88	88
74–84	0–3	81	82	82	83	84	84	84	85	85	85	85	86	86	86	86	86	87	87	87	87
	4–13	82	82	82	83	84	84	84	84	85	85	85	85	85	85	86	86	86	86	86	86
	14+	82	82	82	83	84	84	84	84	84	85	85	85	85	85	85	85	85	85	85	86
85–93	0–3	80	81	81	81	82	82	82	82	82	83	83	83	83	83	83	84	84	84	84	84
	4–13	80	81	81	81	81	81	82	82	82	82	82	82	82	83	83	83	83	83	83	83
	14+	80	81	81	81	81	81	81	81	82	82	82	82	82	82	82	82	82	82	82	83
94–100	0–3	77	77	78	78	78	78	78	79	79	79	79	79	79	80	80	80	80	80	80	81
	4–13	77	77	77	78	78	78	78	78	78	78	78	79	79	79	79	79	79	79	79	79
	14+	77	77	77	77	77	77	77	78	78	78	78	78	78	78	78	78	78	78	78	79

Red values indicate fine fuels are drying. Black values indicate fine fuels are stable or wetting.

Table 3.4.

FFMC daily 25.5–30 °C

Yesterday's FFMC

RH (%)	Wind (km/h)	80	81	82	83	84	85	86	87	88	89	90	91	92	93	94	95	96	97	98	99
0–10	0–3	95	96	96	96	96	96	96	97	97	97	97	97	97	98	98	98	98	98	98	99
	4–13	96	97	97	97	97	97	97	97	97	97	98	98	98	98	98	98	98	98	98	99
	14+	97	97	97	97	97	98	98	98	98	98	98	98	98	98	98	98	98	98	98	99
11–18	0–3	93	93	94	94	94	94	94	95	95	95	95	95	95	96	96	96	96	97	98	98
	4–13	94	94	94	95	95	95	95	95	95	95	95	95	96	96	96	96	96	97	98	98
	14+	95	95	95	95	95	95	95	95	95	96	96	96	96	96	96	96	96	97	98	98
19–28	0–3	91	92	92	92	92	92	93	93	93	93	93	93	94	94	94	95	96	96	96	97
	4–13	92	92	92	93	93	93	93	93	93	93	93	94	94	94	94	95	95	96	96	96
	14+	93	93	93	93	93	93	93	93	93	94	94	94	94	94	94	95	95	96	96	96
29–38	0–3	89	90	90	90	90	90	91	91	91	91	91	92	92	93	94	94	94	94	95	95
	4–13	90	90	90	91	91	91	91	91	91	91	92	92	92	93	93	94	94	94	94	95
	14+	91	91	91	91	91	91	91	91	91	92	92	92	92	93	93	94	94	94	94	94
39–49	0–3	88	88	88	88	89	89	89	89	89	90	90	91	92	92	92	92	93	93	93	93
	4–13	88	88	89	89	89	89	89	89	90	90	90	91	91	92	92	92	92	92	92	93
	14+	89	89	89	89	89	89	89	90	90	90	90	91	91	92	92	92	92	92	92	92
50–61	0–3	86	86	86	86	87	87	87	88	88	89	89	90	90	90	90	91	91	91	91	91
	4–13	86	87	87	87	87	87	87	88	88	89	89	90	90	90	90	90	90	90	91	91
	14+	87	87	87	87	87	87	88	88	88	89	89	89	90	90	90	90	90	90	90	90
62–73	0–3	84	84	84	85	85	85	86	87	87	88	88	88	88	88	88	89	89	89	89	89
	4–13	84	85	85	85	85	86	86	87	87	87	88	88	88	88	88	88	88	88	88	89
	14+	85	85	85	85	85	86	86	87	87	87	87	88	88	88	88	88	88	88	88	88
74–84	0–3	82	82	83	83	84	85	85	85	85	85	86	86	86	86	86	86	87	87	87	87
	4–13	82	83	83	83	84	85	85	85	85	85	85	85	86	86	86	86	86	86	86	86
	14+	83	83	83	83	84	85	85	85	85	85	85	85	85	85	85	85	86	86	86	86
85–93	0–3	80	81	82	82	82	82	82	82	83	83	83	83	83	83	83	84	84	84	84	84
	4–13	80	81	82	82	82	82	82	82	82	82	82	83	83	83	83	83	83	83	83	83
	14+	80	81	82	82	82	82	82	82	82	82	82	82	82	82	82	82	83	83	83	83
94–100	0–3	78	78	78	78	78	78	79	79	79	79	79	79	79	79	80	80	80	80	80	80
	4–13	78	78	78	78	78	78	78	78	78	78	78	79	79	79	79	79	79	79	79	79
	14+	77	78	78	78	78	78	78	78	78	78	78	78	78	78	78	78	78	78	78	79

Red values indicate fine fuels are drying. Black values indicate fine fuels are stable or wetting.

Table 3.5.

FFMC daily

30.5–35.5 °C

Yesterday's FFMC

RH (%)	Wind (km/h)	80	81	82	83	84	85	86	87	88	89	90	91	92	93	94	95	96	97	98	99
0–10	0–3	97	97	97	97	97	97	97	98	98	98	98	98	98	98	98	99	99	99	99	99
	4–13	98	98	98	98	98	98	98	98	98	98	98	98	98	99	99	99	99	99	99	99
	14+	98	98	98	98	98	98	98	98	98	99	99	99	99	99	99	99	99	99	99	99
11–18	0–3	95	95	95	95	95	95	96	96	96	96	96	96	96	96	96	97	97	97	98	98
	4–13	96	96	96	96	96	96	96	96	96	96	96	96	96	97	97	97	97	97	98	98
	14+	96	96	96	96	96	96	96	96	96	96	96	97	97	97	97	97	97	97	98	98
19–28	0–3	93	93	93	93	93	94	94	94	94	94	94	94	94	95	95	95	96	96	97	97
	4–13	94	94	94	94	94	94	94	94	94	94	94	94	95	95	95	95	96	96	97	97
	14+	94	94	94	94	94	94	94	94	94	94	95	95	95	95	95	95	96	96	96	97
29–38	0–3	91	91	91	91	91	92	92	92	92	92	92	92	93	93	94	94	95	95	95	95
	4–13	92	92	92	92	92	92	92	92	92	92	92	93	93	93	94	94	94	95	95	95
	14+	92	92	92	92	92	92	92	92	92	92	93	93	93	93	94	94	94	94	95	95
38–49	0–3	89	89	89	89	90	90	90	90	90	90	91	91	92	92	93	93	93	93	93	94
	4–13	90	90	90	90	90	90	90	90	90	90	91	91	92	92	92	93	93	93	93	93
	14+	90	90	90	90	90	90	90	90	90	91	91	91	92	92	92	92	92	93	93	93
50–61	0–3	87	87	87	87	88	88	88	88	88	89	90	90	90	91	91	91	91	91	91	92
	4–13	87	88	88	88	88	88	88	88	89	89	90	90	90	90	91	91	91	91	91	91
	14+	88	88	88	88	88	88	88	88	89	89	90	90	90	90	90	90	91	91	91	91
62–73	0–3	85	85	85	86	86	86	86	87	88	88	88	88	89	89	89	89	89	89	89	89
	4–13	85	86	86	86	86	86	86	87	88	88	88	88	88	88	89	89	89	89	89	89
	14+	86	86	86	86	86	86	86	87	88	88	88	88	88	88	88	88	88	88	88	89
74–84	0–3	82	83	83	84	84	85	86	86	86	86	86	86	86	86	87	87	87	87	87	87
	4–13	83	83	83	84	84	85	86	86	86	86	86	86	86	86	86	86	86	86	86	86
	14+	83	84	84	84	84	85	86	86	86	86	86	86	86	86	86	86	86	86	86	86
85–93	0–3	80	81	82	82	83	83	83	83	83	83	83	83	83	83	84	84	84	84	84	84
	4–13	80	81	82	82	83	83	83	83	83	83	83	83	83	83	83	83	83	83	83	83
	14+	81	81	82	82	82	83	83	83	83	83	83	83	83	83	83	83	83	83	83	83
94–100	0–3	78	78	78	79	79	79	79	79	79	79	79	79	79	79	79	80	80	80	80	80
	4–13	78	78	78	78	78	78	79	79	79	79	79	79	79	79	79	79	79	79	79	79
	14+	78	78	78	78	78	78	78	78	78	78	78	78	78	79	79	79	79	79	79	79

Red values indicate fine fuels are drying. Black values indicate fine fuels are stable or wetting.

Table 3.6.

FFMC daily   >35.5 °C

Yesterday's FFMC

RH (%)	Wind (km/h)	80	81	82	83	84	85	86	87	88	89	90	91	92	93	94	95	96	97	98	99
0–10	0–3	98	98	98	98	98	98	98	99	99	99	99	99	99	99	99	99	99	99	99	99
	4–3	99	99	99	99	99	99	99	99	99	99	99	99	99	99	99	99	99	99	99	99
	14+	99	99	99	99	99	99	99	99	99	99	99	99	99	99	99	99	99	99	99	99
11–18	0–3	96	96	96	96	97	97	97	97	97	97	97	97	97	97	97	97	97	97	98	99
	4–13	97	97	97	97	97	97	97	97	97	97	97	97	97	97	97	97	97	97	98	99
	14+	97	97	97	97	97	97	97	97	97	97	97	97	97	97	97	97	97	97	98	99
19–28	0–3	94	94	94	95	95	95	95	95	95	95	95	95	95	95	95	96	96	97	97	98
	4–13	95	95	95	95	95	95	95	95	95	95	95	95	95	95	96	96	96	97	97	97
	14+	95	95	95	95	95	95	95	95	95	95	95	95	95	96	96	96	96	97	97	97
29–38	0–3	92	92	92	93	93	93	93	93	93	93	93	93	93	93	94	95	95	95	96	96
	4–13	93	93	93	93	93	93	93	93	93	93	93	93	93	93	94	95	95	95	95	95
	14+	93	93	93	93	93	93	93	93	93	93	93	93	93	93	94	95	95	95	95	95
38–49	0–3	90	90	90	91	91	91	91	91	91	91	91	91	92	93	93	93	94	94	94	94
	4–13	91	91	91	91	91	91	91	91	91	91	91	91	92	93	93	93	93	93	93	94
	14+	91	91	91	91	91	91	91	91	91	91	91	91	92	93	93	93	93	93	93	93
50–61	0–3	88	88	88	88	88	89	89	89	89	89	90	91	91	91	91	91	92	92	92	92
	4–13	89	89	89	89	89	89	89	89	89	89	90	91	91	91	91	91	91	91	91	92
	14+	89	89	89	89	89	89	89	89	89	89	90	91	91	91	91	91	91	91	91	91
62–73	0–3	86	86	86	86	87	87	87	87	88	89	89	89	89	89	89	89	89	90	90	90
	4–13	86	87	87	87	87	87	87	87	88	89	89	89	89	89	89	89	89	89	89	89
	14+	87	87	87	87	87	87	87	87	88	89	89	89	89	89	89	89	89	89	89	89
74–84	0–3	83	84	84	84	85	85	86	86	87	87	87	87	87	87	87	87	87	87	87	87
	4–13	84	84	84	84	85	85	86	86	86	87	87	87	87	87	87	87	87	87	87	87
	14+	84	84	84	85	85	85	86	86	86	86	86	87	87	87	87	87	87	87	87	87
85–93	0–3	81	81	82	83	83	83	83	83	83	84	84	84	84	84	84	84	84	84	84	84
	4–13	81	81	82	83	83	83	83	83	83	83	83	83	83	84	84	84	84	84	84	84
	14+	81	81	82	83	83	83	83	83	83	83	83	83	83	83	83	83	83	83	83	83
94–100	0–3	79	79	79	79	79	79	79	79	79	79	79	79	80	80	80	80	80	80	80	80
	4–13	79	79	79	79	79	79	79	79	79	79	79	79	79	79	79	79	79	79	79	79
	14+	79	79	79	79	79	79	79	79	79	79	79	79	79	79	79	79	79	79	79	79

Red values indicate fine fuels are drying. Black values indicate fine fuels are stable or wetting.

Table 4.1.

FFMC diurnal afternoon & overnight

				Daily FFMC	Local daylight time (h)													
1300	1400	1500	1600	1700	1800	1900	2000	2100	2200	2300	2400	0100	0200	0300	0400	0500	0600	0700
41	43	46	48	50	51	52	53	53	52	51	50	49	48	47	46	45	44	43
48	52	55	57	60	61	62	62	62	61	59	58	56	55	54	52	51	50	49
57	61	65	68	70	70	71	70	69	68	66	65	63	62	60	59	58	56	55
59	63	67	70	72	72	72	72	71	69	68	66	65	63	62	61	59	58	57
62	66	70	72	74	74	74	73	72	71	69	68	66	65	63	62	61	59	58
63	67	71	73	75	75	75	74	73	72	70	69	67	66	64	63	61	60	59
64	68	72	74	76	76	76	75	74	72	71	69	68	66	65	64	62	61	60
66	69	73	75	77	77	77	76	75	73	72	70	69	67	66	64	63	62	60
67	71	75	76	78	78	78	77	76	74	72	71	69	68	67	65	64	63	61
69	72	76	78	79	79	78	77	76	75	73	72	70	69	67	66	65	63	62
71	74	77	79	80	80	79	78	77	76	74	73	71	70	68	67	66	64	63
74	76	79	80	81	81	80	79	78	77	75	73	72	71	69	68	66	65	64
76	78	80	81	82	82	81	80	79	77	76	74	73	71	70	69	67	66	65
78	80	81	82	83	83	82	81	80	78	77	75	74	72	71	70	68	67	66
80	81	82	83	84	84	83	82	81	79	78	76	75	73	72	70	69	68	67
82	82	83	84	85	85	84	83	82	80	79	77	76	74	73	71	70	69	68
83	84	85	85	86	86	85	84	83	81	79	78	77	75	74	72	71	70	68
84	85	86	86	87	87	86	85	83	82	80	79	78	76	75	73	72	71	70
85	86	87	87	88	88	87	86	84	83	81	80	79	77	76	74	73	72	71
86	87	88	89	89	89	88	87	85	84	82	81	80	78	77	75	74	73	72
88	88	89	90	90	90	89	88	86	85	83	82	81	79	78	77	75	74	73
89	89	90	91	91	91	90	89	87	86	84	83	82	80	79	78	76	75	74
90	90	91	92	92	92	91	90	88	87	85	84	83	81	80	79	77	76	75
91	91	92	93	93	93	92	91	89	88	86	85	84	82	81	80	79	77	76
92	93	93	94	94	94	93	92	90	89	88	86	85	84	82	81	80	79	77
93	94	94	95	95	95	94	93	91	90	89	87	86	85	83	82	81	80	79
94	95	95	96	96	96	95	94	92	91	90	88	87	86	85	84	82	81	80
95	96	96	97	97	97	96	95	93	92	91	90	88	87	86	85	84	83	81
96	97	97	98	98	98	97	96	94	93	92	91	90	88	87	86	85	84	83
97	98	98	99	99	99	98	97	95	94	93	92	91	90	88	87	86	85	84
98	99	99	100	100	100	99	98	96	95	94	93	92	91	90	89	88	86	85
1200	1300	1400	1500	1600	1700	1800	1900	2000	2100	2200	2300	2400	0100	0200	0300	0400	0500	0600

Local standard time (h)

To estimate the FFMC during the afternoon or overnight, find today's FFMC in the Daily FFMC column, then move across the row to the column that corresponds to the prediction time.

Table 4.2.
FFMC diurnal   morning

Local daylight time (h)

RH (%)	0700 <68	0700 68–87	0700 >87	0800 <58	0800 58–77	0800 >77	0900 <48	0900 48–67	0900 >67	1000 <43	1000 43–62	1000 >62	1100 <38	1100 38–57	1100 >57	1200 <35	1200 35–54	1200 >54	1300 <33	1300 33–52	1300 >52
50	54	48	43	56	49	44	59	50	45	64	56	51	70	62	57	76	69	64	82	77	72
60	57	53	49	60	54	49	63	55	50	67	60	56	72	66	61	78	0	68	83	79	75
70	62	58	55	64	60	56	67	61	57	71	66	62	76	71	67	80	72	72	85	82	78
72	63	60	57	65	61	57	68	63	58	72	67	63	77	72	68	81	76	73	86	83	79
74	64	61	58	67	63	59	69	64	60	73	69	64	77	73	69	82	77	74	87	84	80
75	65	62	59	67	63	60	70	65	61	74	69	65	78	74	70	82	78	75	87	84	80
76	66	63	60	68	64	60	70	66	61	74	70	66	78	75	70	83	79	76	87	84	81
77	66	63	60	69	65	61	71	67	62	75	71	66	79	75	71	84	80	76	88	85	81
78	67	64	61	69	66	62	72	68	63	75	72	67	79	76	72	84	80	77	88	85	81
79	68	65	62	70	67	63	72	68	64	76	72	68	80	77	72	85	81	77	88	86	82
80	69	66	63	71	67	64	73	69	65	77	73	69	81	77	73	85	82	78	89	86	82
81	69	66	64	72	68	65	74	70	66	77	74	70	81	78	74	86	82	78	89	86	82
82	70	67	65	72	69	66	75	71	67	78	75	70	82	79	74	86	83	79	89	87	83
83	71	68	66	73	70	67	75	72	68	79	76	71	82	79	75	87	83	80	90	87	83
84	72	69	67	74	71	68	76	73	69	80	77	72	83	80	76	87	84	80	90	88	83
85	73	70	67	75	72	69	77	74	70	80	77	73	84	81	77	88	85	81	90	88	83
86	74	71	68	76	73	70	78	75	71	81	78	74	85	82	77	89	86	81	91	88	84
87	75	72	69	77	74	71	79	76	72	82	79	75	85	83	78	89	87	82	91	89	84
88	76	73	71	78	75	72	80	77	73	83	80	76	86	84	79	90	87	82	91	89	84
89	78	74	72	79	76	73	81	78	74	84	81	77	87	85	80	90	88	83	92	89	85
90	79	75	73	80	77	74	82	79	76	85	82	78	88	86	81	91	88	84	92	90	85
91	80	76	74	81	78	75	83	81	77	86	83	79	89	87	82	91	89	84	92	90	85
92	81	77	75	83	79	77	84	82	78	87	85	80	90	88	83	92	89	85	92	90	86
93	83	78	76	84	81	78	85	83	79	88	86	82	91	89	84	92	90	85	93	91	86
94	84	80	77	85	82	79	86	84	81	89	87	83	92	90	85	93	91	86	93	91	86
95	86	81	79	87	83	81	88	85	82	90	88	84	93	91	86	93	91	86	93	91	86
96	87	82	80	88	84	82	89	87	84	91	89	85	94	92	87	94	92	87	94	92	87
97	89	84	81	90	86	83	90	88	85	93	91	87	95	93	88	95	93	88	95	93	88
98	90	85	83	91	87	85	92	89	87	94	92	88	96	94	89	96	94	89	96	94	89
99	92	87	84	93	88	86	93	90	88	95	93	90	97	96	91	97	96	91	97	96	90
100	93	88	85	94	90	88	95	91	90	96	94	91	98	97	92	98	97	92	98	97	92

Yesterday's Standard Daily FFMC

| 0600 | 0700 | 0800 | 0900 | 1000 | 1100 | 1200 |

Local standard time (h)

To estimate the FFMC during the morning, find Yesterday's Standard Daily FFMC, then move across the row to the column with the estimated RH for the prediction time.

Table 4.3.

FFMC slope aspect adjustment

Ground slope (%) and aspect

Level	1%–15%				16%–30%				31%–45%				46%–60%			
	N	E	S	W	N	E	S	W	N	E	S	W	N	E	S	W
80	78	79	82	80	77	78	82	80	74	77	83	81	72	76	84	81
82	80	81	84	82	76	80	84	82	76	79	85	83	74	78	85	83
84	83	83	85	84	79	82	86	84	79	81	87	84	76	80	88	84
86	85	85	87	86	83	84	88	86	81	83	89	86	78	82	90	86
87	86	86	88	87	84	85	89	87	82	84	90	87	80	83	90	87
88	87	87	89	88	85	87	90	88	83	86	91	88	82	85	91	88
89	88	88	90	89	87	88	91	89	85	87	91	89	83	86	92	89
90	89	89	91	90	88	89	92	90	86	88	92	90	84	87	93	90
91	90	90	92	91	89	90	92	91	87	89	93	91	86	88	93	91
92	91	91	93	92	90	91	93	92	88	90	94	92	87	89	94	92
93	92	92	94	93	91	92	94	93	89	91	95	93	88	90	95	93
94	93	93	95	94	92	93	95	94	91	92	96	94	90	92	96	94
	N	E	S	W	N	E	S	W	N	E	S	W	N	E	S	W
	1°–8.5°				9°–17°				18°–24°				25°–31°			

Ground slope (°) and aspect

These adjustments should only be applied in open fuel types on clear days in March, April, August, September, or October between 1200 and 2000 local standard time. To use this table, find the FFMC, determine the slope and aspect of the weather observation point, and obtain the corresponding adjusted FFMC value for open fuel types.

Table 5.

Slope equivalent wind speed (km/h)

Fuel type		Ground slope (%)						
		10%	20%	30%	40%	50%	60%	70%
C-1		1	3	4	6	8	10	12
C-2		3	7	12	17	23	29	36
C-3		2	4	6	9	12	15	18
C-4		3	7	12	17	23	29	37
C-5		1	3	5	7	9	12	14
C-6		2	4	7	10	13	17	21
C-7		2	5	9	13	17	21	26
D-1, D-2		3	7	11	16	21	26	32
M-1, M-2	50% PC	3	7	11	16	22	28	35
M-3	60% PDF	3	7	12	18	25	34	120
M-4	60% PDF	3	12	12	17	23	30	38
O-1a		3	8	13	19	25	32	41
O-1b		3	7	11	15	21	26	33
S-1		4	8	14	20	27	35	45
S-2		3	7	11	16	21	26	35
S-3		2	4	6	9	12	16	20
		6°	11°	17°	22°	27°	31°	35°
				Ground slope (°)				

Note: Values are for FFMC 90 and are within +/- 2 km/h for FFMC 80–96, except the values for FFMC > 94 and slope > 50 may be underestimated by 5+ km/h. For slopes > 70% use the 70% value.

Table 6.1.

Beaufort scale
for estimating 10-m open wind speed

Wind speed (km/h)			
Range	Mean	Description	Observed wind effects
<1	0	Calm	Smoke rises vertically
1–5	3	Light air	Direction of drift shown by smoke drift but not by wind vanes
6–11	9	Light breeze	Wind felt on face, leaves rustle; vanes moved by wind
12–19	16	Gentle breeze	Leaves and twigs in constant motion; wind extends light flag
20–28	24	Moderate breeze	Raises dust and loose paper, small branches are moved
29–38	34	Fresh breeze	Small trees in leaf begin to sway; crested wavelets on inland waters
39–49	44	Strong breeze	Large branches in motion, whistling in telephone wires, umbrellas used with difficulty
50–61	55	Near gale	Whole trees in motion; inconvenience felt when walking against wind
62–74	68	Moderate gale	Breaks twigs off trees, generally impedes progress
75–88	82	Strong gale	Slight structural damage occurs
89–102	96	Whole gale	Seldom experienced inland, trees uprooted; considerable structural damage
103–117	110	Storm	Very rarely experienced; widespread damage
118+	125	Hurricane	Severe widespread damage to vegetation and significant structural damage possible

Adapted from List, R.J. 1951. Smithsonian meteorological tables. 6th Rev. Ed. Smithsonian Inst. Press, Washington, DC.

Table 6.2.

Wind speed adjustment factor to 10-m height

Measured height (m)	Rough surface	Smooth surface
1.5	1.94	1.48
2.0–2.9	1.54	1.31
3.0–3.9	1.37	1.22
4.0–4.9	1.26	1.16
5.0–6.9	1.18	1.11
7.0–8.9	1.06	1.03

Multiply the wind speed at the measured height by the correction factor to get the 10-m wind speed.

Table 6.3.

ISI Initial Spread Index / BISI

Effective wind speed (km/h)

FFMC	0	5	10	15	20	25	30	35	40	45	50
77	1	1	1	2	2	3	4	5	7	9	9
	1	1	1	0	0	0	0	0	0	0	0
78	1	1	2	2	3	3	4	5	7	8	9
	1	1	1	0	0	0	0	0	0	0	0
79	1	1	2	2	3	4	5	6	8	9	10
	1	1	1	1	0	0	0	0	0	0	0
80	1	1	2	2	3	4	5	7	9	10	11
	1	1	1	1	0	0	0	0	0	0	0
81	1	2	2	3	3	4	6	7	10	11	13
	1	1	1	1	0	0	0	0	0	0	0
82	1	2	2	3	4	5	6	8	11	13	14
	1	1	1	1	1	0	0	0	0	0	0
83	2	2	3	3	4	6	7	9	12	15	16
	2	1	1	1	1	0	0	0	0	0	0
84	2	2	3	4	5	6	8	11	14	17	18
	2	1	1	1	1	1	0	0	0	0	0
85	2	3	3	4	6	7	10	12	16	19	21
	2	2	1	1	1	1	0	0	0	0	0
86	2	3	4	5	7	9	11	14	18	22	24
	2	2	1	1	1	1	1	0	0	0	0
87	3	4	5	6	8	10	13	16	21	25	28
	3	2	2	1	1	1	1	0	0	0	0
88	3	4	5	7	9	11	15	19	24	29	32
	3	3	2	2	1	1	1	1	0	0	0
89	4	5	6	8	10	13	17	22	28	33	37
	4	3	2	2	1	1	1	1	0	0	0
90	4	6	7	9	12	15	19	25	32	39	43
	4	3	3	2	2	1	1	1	1	0	0
91	5	6	8	11	14	17	22	29	37	45	50
	5	4	3	2	2	1	1	1	1	1	0
92	6	7	9	12	16	20	26	33	43	51	57
	6	4	3	3	2	2	1	1	1	1	0
93	7	8	11	14	18	23	30	38	49	59	66
	7	5	4	3	2	2	1	1	1	1	1
94	8	10	12	16	21	27	34	44	57	68	76
	8	6	5	4	3	2	2	1	1	1	1
95	9	11	14	18	24	31	39	51	65	78	87
	9	7	5	4	3	2	2	1	1	1	1
96	10	13	16	21	27	35	45	58	74	89	99
	10	8	6	5	4	3	2	2	1	1	1
97	11	15	19	24	31	40	51	66	85	102	114
	11	9	7	5	4	3	3	2	2	1	1
98	13	17	21	28	36	46	59	76	97	117	130
	13	10	8	6	5	4	3	2	2	1	1
	0	1.4	2.8	4.2	5.6	6.9	8.3	9.7	11.1	12.5	13.9

Effective wind speed (m/s)

Use red values for the head fire and black values for the backfire.

Table 7.1.

DMC Duff Moisture Code daily drying factors

Temperature (°C)	Relative humidity (%)	Month						
		Apr	May	Jun	Jul	Aug	Sep	Oct
10.5–15.0	0–42	3	3	3	3	2	2	2
	43–73	1	2	2	1	1	1	1
	74–100	0	0	0	0	0	0	0
15.5–20.0	0–32	4	4	4	4	3	3	2
	33–52	3	3	3	3	2	2	2
	53–73	2	2	2	2	1	1	1
	74–100	1	1	1	1	0	0	0
20.5–25.0	0–32	5	5	5	5	4	3	3
	33–52	3	3	3	3	3	2	2
	53–73	2	2	2	2	2	2	1
	74–100	1	1	1	1	1	1	0
25.5–30.0	0–32	6	6	6	6	5	4	3
	33–52	4	4	4	4	3	3	3
	53–73	3	3	3	2	2	2	2
	74–100	1	1	1	1	1	1	1
30.5–35.0	0–32	7	7	7	7	6	5	4
	33–52	5	5	5	5	4	3	3
	53–73	3	3	3	3	3	2	2
	74–100	1	1	1	1	1	1	1

Table 7.2.

DC Drought Code daily drying factors

Temperature (°C)	Month						
	Apr	May	Jun	Jul	Aug	Sep	Oct
10.5–15.0	3	5	6	6	5	4	3
15.5–20.0	4	6	7	7	6	5	4
20.5–25.0	5	6	7	8	7	6	5
25.5–30.0	6	7	8	9	8	7	6
30.5–35.0	7	8	9	10	9	8	7

Table 8.1.

BUI Buildup Index

	DC										
DMC	0–19	20–39	40–59	60–79	80–99	100–119	120–139	140–159	160–179	180–199	200–224
11	11	11	14	16	17	18	18	19	19	19	19
12	12	12	15	17	18	19	19	20	20	21	21
13	13	13	16	18	19	20	21	21	22	22	23
14	13	14	16	19	20	21	22	23	23	24	24
15–16	15	15	17	20	22	23	24	25	25	26	26
17–18	17	17	19	21	24	25	26	27	28	28	29
19–20	19	19	20	23	25	27	28	29	30	31	32
21–22	21	21	21	24	27	29	30	32	33	33	34
23–24	23	23	23	25	28	31	32	34	35	36	37
25–27	25	26	26	27	30	33	35	36	38	39	40
28–30	28	29	29	29	32	35	37	39	41	42	43
31–33	31	32	32	32	34	37	40	42	43	45	46
34–36	34	34	35	35	35	39	42	44	46	48	50
37–39	37	37	38	38	38	41	44	46	49	51	52
40–43	41	41	41	41	41	43	46	49	51	54	56
44–47	44	45	45	45	45	45	48	52	54	57	59
48–51	48	49	49	49	49	49	51	54	57	60	63
52–55	52	53	53	53	53	53	53	56	60	63	66
56–60	57	57	57	58	58	58	58	59	63	66	69
61–65	62	62	62	62	63	63	63	63	65	69	72
66–70	67	67	67	67	68	68	68	68	68	72	75
71–75	72	72	72	72	72	73	73	73	73	74	78
76–81	77	77	77	78	78	78	78	78	78	78	82
82–87	83	83	83	84	84	84	84	84	84	84	85
88–93	89	89	89	89	90	90	90	90	90	90	90
94–100	95	95	96	96	96	96	96	97	97	97	97
101–107	102	102	102	103	103	103	103	103	104	104	104
108–115	109	110	110	110	110	110	111	111	111	111	111
116–124	118	118	118	118	119	119	119	119	119	119	120
125–134	127	127	127	128	128	128	128	128	129	129	129
135–145	137	137	138	138	138	138	139	139	139	139	139
146–157	148	149	149	149	149	150	150	150	150	150	151
158–170	160	161	161	161	162	162	162	162	162	163	163
171–186	174	175	175	175	176	176	176	176	177	177	177
187–205	191	192	192	192	193	193	193	193	194	194	194

Table 8.2.

BUI Buildup Index

DMC	DC									
	225–249	250–274	275–299	300–329	330–359	360–399	400–439	440–489	490–539	540–599
11	20	20	20	20	20	20	21	21	21	21
12	21	22	22	22	22	22	22	23	23	23
13	23	23	23	24	24	24	24	24	24	25
14	24	25	25	25	25	26	26	26	26	26
15–16	27	27	27	28	28	28	28	29	29	29
17–18	30	30	30	31	31	31	32	32	32	33
19–20	32	33	33	34	34	35	35	35	36	36
21–22	35	36	36	37	37	38	38	39	39	39
23–24	38	38	39	40	40	41	41	42	42	43
25–27	41	42	42	43	44	44	45	46	46	47
28–30	44	45	46	47	48	49	49	50	51	51
31–33	48	49	50	51	52	53	54	55	55	56
34–36	51	52	54	55	56	57	58	59	60	61
37–39	54	56	57	58	60	61	62	63	64	65
40–43	58	59	61	62	64	65	67	68	69	70
44–47	61	63	65	67	68	70	72	73	75	76
48–51	65	67	69	71	73	74	76	78	80	81
52–55	68	71	73	75	77	79	81	83	85	87
56–60	72	75	77	79	82	84	86	88	90	92
61–65	76	79	81	84	86	89	92	94	96	99
66–70	79	82	85	88	91	94	97	100	102	105
71–75	82	86	89	92	95	98	102	105	108	111
76–81	86	90	93	97	100	103	107	110	114	117
82–87	89	94	97	101	105	108	112	116	120	123
88–93	93	97	101	105	109	113	118	122	126	130
94–100	97	101	105	110	114	118	123	127	132	136
101–107	104	104	109	114	119	123	128	133	138	143
108–115	111	111	113	118	123	128	134	139	145	150
116–124	120	120	120	123	128	133	140	146	152	157
125–134	129	129	129	129	134	139	146	153	159	165
135–145	139	140	140	140	140	145	153	160	167	173
146–157	151	151	151	151	151	151	159	167	175	182
158–170	163	163	163	164	164	164	166	174	183	191
171–186	177	177	178	178	178	178	178	182	191	200
187–205	194	195	195	195	195	195	196	196	201	211

Table 9.1.
Equilibrium ROS (m/min)
Fire Intensity Class

C-1 spruce–lichen woodland

Intensity class

1	< 10 kW/m
2	10–500
3	500–2 000
4	2 000–4 000
5	4 000–10 000
6	>10 000

ISI	0–20	21–30	31–40	41–60	61–80	81–120	121–160	161–200
1	0	0	0	0	0 ①	0	0	0
2	0	0	0	0	0	0	0	0
3	<0.1	<0.1	<0.1	<0.1	<0.1 ②	<0.1	<0.1	<0.1
4	<0.1	0.1	0.1	0.1	0.1	0.1	0.1	0.1
5	0.2	0.2	0.3	0.3	0.3	0.3	0.3	0.3
6	0.4	0.5	0.5	0.5	0.6	0.6	0.6	0.6
7	0.6	0.8	0.9	0.9	1	1	1	1
8	1	1	1	1	2	2	2	2
9	1	2*	2*	2*	2* ③	2*	2*	2*
10	2*	3*	3*	3*	3*	3*	3*	3*
11	3*	4*	4*	4*	4*	4*	5*	5*
12	4*	5*	5*	5*	6* ④	6*	6*	6*
13	5*	6*	7*	7*	7*	7*	7*	7*
14	6*	8*	8*	9*	9* ⑤	9*	9*	9*
15	7*	9*	10*	10*	11*	11*	11*	11*
16	8*	11*	12	12	13	13	13	13
17	9*	13	14	14	15	15	15	15
18	11*	15	16	16	17	17	17	18
19	12	17	18	19	19 ⑥	20	20	20
20	14	19	20	21	21	22	22	22
21–25	18	25	27	28	29	29	30	30
26–30	26	35	38	39	40	41	42	42
31–35	33	45	48	50	51	52	53	54
36–40	39	53	56	59	60	62	63	63
41–45	44	59	63	66	68	69	70	71
46–50	47	64	68	71	73	75	76	77
51–55	50	68	72	75	78	79	81	81
56–60	52	71	75	78	81	83	84	85
61–65	54	73	77	81	83	85	86	87
66–70	55	75	79	83	85	87	88	89

Constant values: foliar moisture content = 97%; CBH = 2 m; surface fuel consumption for FFMC 90. □ = average BUI. Type of fire: Black values = surface with <10% CFB, black values with * = intermittent crown with 10–89% CFB, white values = **continuous crown fire** , ▁ = approximately 50% CFB value. ○ = intensity class.

Table 9.2.
Equilibrium ROS (m/min)
Fire Intensity Class

C-2 boreal spruce

Intensity class
- 1 < 10 kW/m
- 2 10–500
- 3 500–2 000
- 4 2 000–4 000
- 5 4 000–10 000
- 6 >10 000

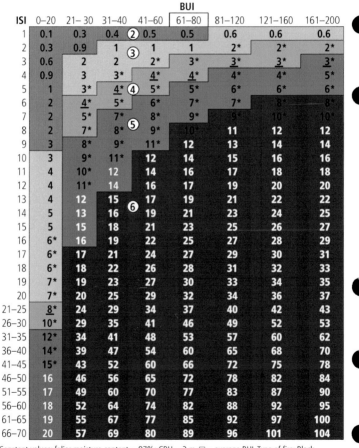

ISI	0–20	21–30	31–40	41–60	61–80	81–120	121–160	161–200
1	0.1	0.3	0.4 ②	0.5	0.5	0.6	0.6	0.6
2	0.3	0.9	1 ③	1	1	2*	2*	2*
3	0.6	2	2	2*	3*	<u>3*</u>	<u>3*</u>	<u>3*</u>
4	0.9	3	3*	<u>4*</u>	<u>4*</u>	4*	4*	5*
5	1	3*	<u>4*</u> ④	5*	5*	6*	6*	6*
6	2	<u>4*</u>	5*	6*	7*	7*	8*	8*
7	2	5*	7*	8*	9*	9*	10*	10*
8	2	7*	8* ⑤	9*	10*	11	12	12
9	3	8*	9*	11*	12	13	14	14
10	3	9*	11*	12	14	15	16	16
11	4	10*	12	14	16	17	18	18
12	4	11*	14	16	17	19	20	20
13	4	12	15	17	19	21	22	22
14	5	13	16 ⑥	19	21	23	24	25
15	5	15	18	21	23	25	26	27
16	6*	16	19	22	25	27	28	29
17	6*	17	21	24	27	29	30	31
18	6*	18	22	26	28	31	32	33
19	7*	19	23	27	30	33	34	35
20	7*	20	25	29	32	34	36	37
21–25	<u>8*</u>	24	29	34	37	40	42	43
26–30	10*	29	35	41	46	49	52	53
31–35	12*	34	41	48	53	57	60	62
36–40	14*	39	47	54	60	65	68	70
41–45	15*	43	52	60	66	72	75	78
46–50	16	46	56	65	72	78	82	84
51–55	17	49	60	70	77	83	87	90
56–60	18	52	64	74	82	88	92	95
61–65	19	55	67	77	85	92	97	100
66–70	20	57	69	80	89	96	101	104

Constant values: foliar moisture content = 97%; CBH = 3 m. □ = average BUI. Type of fire: Black values = surface with <10% CFB, black values with * = intermittent crown with 10–89% CFB, white values = **continuous crown fire** , — = approximately 50% CFB value. ◯ = intensity class.

Table 9.3.
Equilibrium ROS (m/min)
Fire Intensity Class

C-3 mature jack or lodgepole pine

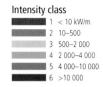

Intensity class
1 < 10 kW/m
2 10–500
3 500–2 000
4 2 000–4 000
5 4 000–10 000
6 >10 000

				BUI				
ISI	0–20	21–30	31–40	41–60	61–80	81–120	121–160	161–200
1	0	0	0	0	0	① 0	<0.1	<0.1
2	<0.1	<0.1	<0.1	<0.1	<0.1	<0.1	<0.1	<0.1
3	<0.1	0.2	0.2	0.2	0.2	② 0.2	0.2	0.2
4	0.1	0.3	0.4	0.4	0.5	0.5	0.5	0.6
5	0.3	0.6	0.7	0.8	0.9	③ 0.9	1	1
6	0.4	1	1	1	1	2	2	2
7	0.6	2	2	2	2	④ 2	2	2
8	0.9	2	2	3	3	3	3	3*
9	1	3	3	4	4	⑤ 4	4*	5*
10	2	4	4	5	5	6*	6*	6*
11	2	5	5	6	7*	7*	7*	7*
12	2	6	7	7	8*	8*	9*	9*
13	3	7	8	9	10*	⑥ 10*	11*	11*
14	3	8	9	10*	11*	12*	12*	13*
15	4	9	11	12*	13*	14*	14	15
16	4	10	12	14*	15*	16	16	17
17	5	12	14	16*	17	18	19	19
18	6	13	15	17*	19	20	21	21
19	6	15	17*	19	21	22	23	24
20	7	16	19*	21	23	25	26	26
21–25	9	21	24*	27	30	31	33	34
26–30	12	28*	33	38	41	43	45	46
31–35	16	36*	42	47	51	55	57	58
36–40	18	43	50	56	61	65	68	69
41–45	21	49	57	64	70	74	77	79
46–50	23	54	63	71	77	82	86	88
51–55	25	58	69	77	84	89	93	95
56–60	27	62	73	82	89	95	99	101
61–65	28	65	77	86	94	100	104	106
66–70	29	68	80	90	97	103	108	110

Constant values: foliar moisture content = 97%; CBH = 8 m. □ = average BUI. Type of fire: Black values = surface with <10% CFB, black values with * = intermittent crown with 10–89% CFB, white values = **continuous crown fire** , ▬ = approximately 50% CFB value. ○ = intensity class.

Table 9.4.
Equilibrium ROS (m/min)
Fire Intensity Class

C-4 immature jack or lodgepole pine

Intensity class
1 < 10 kW/m
2 10–500
3 500–2 000
4 2 000–4 000
5 4 000–10 000
6 >10 000

					BUI			
ISI	0–20	21–30	31–40	41–60	61–80	81–120	121–160	161–200
1	0.2 ①	0.4	0.5	0.5	0.5	0.6	0.6	0.6
2	0.6	② 1	1	1	2	2	2*	2*
3	1	2	2	3	3*	3*	3*	3*
4	2	3	3	4*	4*	4*	4*	5*
5	2	③ 4	5	5*	6*	6*	6*	6*
6	3	5	6*	7*	7*	8*	8*	8*
7	4	7	8*	8*	9*	9*	10*	10*
8	4	8	9*	10*	11*	11*	11	12
9	5	9	11*	12*	12	13	13	14
10	6	④ 11*	12*	13	14	15	15	16
11	6	12*	14*	15	16	17	17	18
12	7	14*	15*	17	18	19	19	20
13	8	⑤ 15*	17	19	20	21	21	22
14	9	16*	19	20	22	23	23	24
15	9	18*	20	22	24	25	26	26
16	10	19*	22	24	25	27	28	28
17	11	21	23	26	27	29	30	30
18	12	22	25	27	29	31	32	32
19	12	⑥ 23	27	29	31	33	34	34
20	13	25	28	31	33	34	36	36
21–25	15	29	33	36	38	40	41	42
26–30	18	35	40	44	47	49	50	51
31–35	21	41	46	51	54	57	59	60
36–40	24	46	52	57	61	64	66	67
41–45	27	51	58	63	67	71	73	74
46–50	29	55	62	69	73	76	79	80
51–55	31	59	67	73	78	82	84	86
56–60	32	62	70	77	82	86	89	90
61–65	34	65	74	81	86	90	93	95
66–70	35	67	76	84	89	94	97	98

Constant values: foliar moisture content = 97%; CBH = 4 m. □ = average BUI. Type of fire: Black values = surface with <10% CFB, black values with * = intermittent crown with 10–89% CFB, white values = **continuous crown fire** , — = approximately 50% CFB value. ○ = intensity class.

Table 9.5.
Equilibrium ROS (m/min)
Fire Intensity Class

C-5 red and white pine

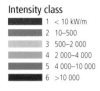

Intensity class
- 1 < 10 kW/m
- 2 10–500
- 3 500–2 000
- 4 2 000–4 000
- 5 4 000–10 000
- 6 >10 000

				BUI				
ISI	0–20	21–30	31–40	41–60	61–80	81–120	121–160	161–200
1	0	0	0	0	0	0	0	0
2	0	0	0	0	0	0	0	0
3	<0.1	<0.1	<0.1	<0.1	<0.1	<0.1	<0.1	<0.1
4	<0.1	<0.1	<0.1	0.1	0.1	0.1	0.1	0.1
5	<0.1	0.2	0.2	0.2	0.2	0.2	0.3	0.3
6	0.2	0.3	0.4	0.4	0.4	0.4	0.5	0.5
7	0.3	0.5	0.6	0.7	0.7	0.7	0.8	0.8
8	0.4	0.8	0.9	1	1	1	1	1
9	0.6	1	1	1	1	2	2	2
10	0.8	2	2	2	2	2	2	2
11	1	2	2	2	3	3	3	3
12	1	2	3	3	3	3	3	4
13	2	3	3	4	4	4	4	4
14	2	4	4	4	5	5	5	5
15	2	4	5	5	6	6	6	6
16	3	5	5	6	6	7	7	7
17	3	5	6	7	7	8	8	8
18	3	6	7	8	8	9	9	9
19	4	7	8	9	9	10	10	10
20	4	8	9	9	10	10	11	11*
21–25	5	10	11	12	13	13	14*	14*
26–30	7	13	14	16	17	18*	18*	19*
31–35	8	15	18	19	20	21*	22	23
36–40	9	18	20	22	23	24*	25	26
41–45	10	19	22	24	25*	27	28	28
46–50	11	20	23	25	27*	28	29	30
51–55	11	21	24	27	28*	30	31	31
56–60	11	22	25	27	29*	31	32	32
61–65	12	22	25	28	30*	31	32	33
66–70	12	23	26	28	30*	32	33	33

Constant values: foliar moisture content = 97%; CBH = 18 m. □ = average BUI. Type of fire: Black values = surface with <10% CFB, black values with * = intermittent crown with 10–89% CFB, white values = **continuous crown fire** , ___ = approximately 50% CFB value. ◯ = intensity class.

Table 9.6.
Equilibrium ROS (m/min)
Fire Intensity Class

C-6 conifer plantation, 7-m CBH

Intensity class
- 1 < 10 kW/m
- 2 10–500
- 3 500–2 000
- 4 2 000–4 000
- 5 4 000–10 000
- 6 >10 000

| | | | | BUI | | | |
ISI	0–20	21–30	31–40	41–60	61–80	81–120	121–160	161–200
1	0	<0.1	<0.1	<0.1	<0.1 ①	<0.1	<0.1	<0.1
2	<0.1	<0.1	<0.1	<0.1	<0.1	0.1	0.1	0.1
3	0.1	0.2	0.3	0.3	0.3 ②	0.3	0.3	0.3
4	0.2	0.5	0.5	0.6	0.6	0.7	0.7	0.7
5	0.4	0.8	0.9	1	1	1	1	1
6	0.7	1	1	2	2	2	2	2
7	1	2	2	2	2	3	3	4
8	1	2	3	3	3 ③	3	6*	7*
9	2	3	4	4	4 ④	7*	10*	11*
10	2	4	4	5	5	11*	14*	15*
11	2	5	5	6	8	15*	18*	19*
12	3	5	6	7	13* ⑤	19*	21*	22*
13	3	6	7	8	17*	22*	24*	24*
14	4	7	8	9	21*	25*	26*	27*
15	4	8	9	10	24*	28*	29*	29*
16	5	9	10	16* ⑥	27*	30*	31*	31
17	5	10	11	20*	30*	32*	33	33
18	5	10	12	24*	32*	34	34	34
19	6	11	13	28*	34*	35	36	36
20	6	12	13	30*	35*	37	37	37
21–25	7	14	16	37*	40	40	40	41
26–30	9	17	22*	43	45	45	45	45
31–35	10	19	34*	47	48	48	48	48
36–40	10	20	41*	50	51	51	51	51
41–45	11	21	45*	52	53	53	53	53
46–50	11	22	48*	54	54	54	54	54
51–55	12	22	50*	55	56	56	56	56
56–60	12	23	51*	56	57	57	57	57
61–65	12	23	52*	57	57	57	57	57
66–70	12	23	53*	58	58	58	58	58

Constant values: foliar moisture content = 97%; CBH = 7 m. □ = average BUI. Type of fire: Black values = surface with <10% CFB, black values with * = intermittent crown with 10–89% CFB, white values = **continuous crown fire**, — = approximately 50% CFB value. ○ = intensity class.

Table 9.7.
Equilibrium ROS (m/min)
Fire Intensity Class

C-6 conifer plantation, 2-m CBH

Intensity class
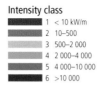
1 < 10 kW/m
2 10–500
3 500–2 000
4 2 000–4 000
5 4 000–10 000
6 >10 000

ISI	0–20	21–30	31–40	41–60	61–80	81–120	121–160	161–200
1	0	<0.1	<0.1	<0.1	① <0.1	<0.1	<0.1	<0.1
2	<0.1	<0.1	<0.1	<0.1	<0.1	0.1	0.1	0.1
3	0.1	0.2	0.3	0.3 ②	0.3	0.3	0.3	0.3
4	0.2	0.5	0.5	0.6	0.6	0.9	1	1
5	0.4	0.8	0.9	1	③ 2	3*	3*	3*
6	0.7	1	1	2	④ 4*	5*	6*	6*
7	1	2	2	5*	7*	8*	9*	9*
8	1	2	3	8*	⑤ 10*	11*	12*	12*
9	2	3	6*	11*	⑥ 13*	14*	15*	15*
10	2	4	10*	15*	17*	17*	18*	18*
11	2	5	14*	18*	20*	20*	21*	21*
12	3	6	17*	21*	22*	23*	23*	24*
13	3	11*	21*	24*	25*	25*	26*	26*
14	4	15*	23*	26*	27*	28*	28*	28*
15	4	19*	26*	28*	29*	30	30	30
16	5	22*	28*	30*	31	31	32	32
17	5	25*	31*	32	33	33	33	33
18	5	27*	32*	34	34	35	35	35
19	6	30*	34*	35	36	36	36	36
20	6	32*	36	37	37	37	37	37
21–25	7	37*	40	40	40	41	41	41
26–30	9	43	44	45	45	45	45	45
31–35	10	47	48	48	48	48	48	48
36–40	10	50	51	51	51	51	51	51
41–45	11	52	53	53	53	53	53	53
46–50	11	54	54	54	54	54	54	54
51–55	12	55	56	56	56	56	56	56
56–60	12	56	56	57	57	57	57	57
61–65	12	57	57	57	57	57	57	57
66–70	12	57	58	58	58	58	58	58

Constant values: foliar moisture content = 97%; CBH = 2 m. □ = average BUI. Type of fire: Black values = surface with <10% CFB, black values with * = intermittent crown with 10–89% CFB, white values = **continuous crown fire** , ▬ = approximately 50% CFB value. ◯ = intensity class.

Table 9.8.
Equilibrium ROS (m/min)
Fire Intensity Class

C-7 ponderosa pine/Douglas-fir

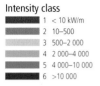

Intensity class
1 < 10 kW/m
2 10–500
3 500–2 000
4 2 000–4 000
5 4 000–10 000
6 >10 000

					BUI			
ISI	0–20	21–30	31–40	41–60	61–80	81–120	121–160	161–200
1	<0.1	<0.1	<0.1	<0.1	<0.1	② <0.1	<0.1	<0.1
2	<0.1	0.1	0.1	0.1	0.2	0.2	0.2	0.2
3	0.2	0.3	0.3	0.3	0.3	0.3	0.4	0.4
4	0.3	0.5	0.5	0.5	0.6	0.6	0.6	0.6
5	0.4	0.7	0.8	0.8	0.9	③ 0.9	0.9	0.9
6	0.6	1	1	1	1	1	1	1
7	0.8	1	1	2	2	2	2	2
8	1	2	2	2	2	2	2	2
9	1	2	2	2	2	3	3	3
10	2	2	3	3	3	④ 3	3	3
11	2	3	3	3	4	4	4	4
12	2	3	4	4	4	4	4	4
13	2	4	4	4	5	5	5	5
14	3	4	5	5	5	⑤ 5	6	6
15	3	5	5	6	6	6	6*	6*
16	3	5	6	6	6*	7*	7*	7*
17	4	6	6	7*	7*	7*	8*	<u>8</u>*
18	4	6	7	7*	8*	8*	<u>8</u>*	8*
19	4	7	7*	8*	8*	<u>9</u>*	9*	9*
20	5	7	8*	<u>9</u>*	9*	9*	10*	10*
21–25	6	9*	<u>10</u>*	11*	11*	⑥ 11*	12*	12*
26–30	7	<u>12</u>*	13*	14*	14*	15*	15*	15*
31–35	9	14*	16*	17	17	18	18	19
36–40	10*	17*	18	19	20	21	22	22
41–45	<u>12</u>*	19	21	22	23	24	24	25
46–50	13*	21	23	24	26	26	27	27
51–55	14*	23	25	27	28	29	29	30
56–60	15*	24	27	28	30	31	32	32
61–65	16*	26	28	30	32	33	33	34
66–70	17*	27	30	32	33	34	35	36

Constant values: foliar moisture content = 97%; CBH = 10 m. forest floor consumption for FFMC 90. □ = average BUI. Type of fire: Black values = surface with <10% CFB, black values with * = intermittent crown with 10–89% CFB, white values = **continuous crown fire** , ─ = approximately 50% CFB value. ◯ = intensity class.

Table 9.9.
Equilibrium ROS (m/min)
Fire Intensity Class

D-1 leafless aspen

Intensity class
1 < 10 kW/m
2 10–500
3 500–2 000
4 2 000–4 000
5 4 000–10 000
6 >10 000

ISI	0–20	21–30	31–40	41–60	61–80	81–120	121–160	161–200
1	<0.1 ①	<0.1	<0.1	<0.1	<0.1	<0.1	<0.1	<0.1
2	0.1	0.2	0.2	0.2	0.2	0.2	0.2	0.2
3	0.3	0.4	0.4	0.4	0.4	0.4	0.5	0.5
4	0.4	0.6	0.6	0.7	0.7	0.7	0.7	0.7
5	0.6	0.8	0.9	0.9	1	1	1	1
6	0.8	1	1	1	1	1	1	1
7	1	② 1	1	2	2	2	2	2
8	1	2	2	2	2	2	2	2
9	1	2	2	2	2	2	2 ③	2
10	2	2	2	3	3	3	3	3
11	2	3	3	3	3	3	3	3
12	2	3	3	3	3	3	4	4
13	2	3	4	4	4	4	4	4
14	3	4	4	4	4	4	4	4
15	3	4	4	4	5	5	5	5
16	3	4	5	5	5	5	5	5
17	4	5	5	5	5	6	6	6 ④
18	4	5	5	6	6	6	6	6
19	4	6	6	6	6	6	7	7
20	4	6	6	7	7	7	7	7
21–25	5	7	7	8	8	8	8	8
26–30	6	9	9	10	10	10	10	11
31–35	8	11	11	12	12	12	13	13
36–40	9	12	13	14	14	14	14	15 ⑤
41–45	10	14	15	15	16	16	16	16
46–50	11	15	16	17	17	18	18	18
51–55	12	17	18	18	19	19	20	20
56–60	13	18	19	20	20	21	21	21
61–65	14	19	20	21	22	22	22	23
66–70	15	20	21	22	23	23	24	24 ⑥

Note: crown fires are not expected in deciduous fuel types but high intensity surface fires can occur. □ = average BUI. Type of fire: surface fire with <10% CFB. ○ = intensity class.

Table 9.10.
Equilibrium ROS (m/min)
Fire Intensity Class

D-2 green aspen

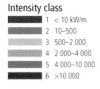

Intensity class
- 1 < 10 kW/m
- 2 10–500
- 3 500–2 000
- 4 2 000–4 000
- 5 4 000–10 000
- 6 >10 000

				BUI			
ISI	0–20	21–30	31–40	41–60	61–80	81–120	121–160	161–200
1	0	0	0	0	<0.1	<0.1	<0.1	<0.1
2	0	0	0	0	<0.1	<0.1	<0.1	<0.1
3	0	0	0	0	<0.1	<0.1	<0.1	<0.1
4	0	0 ①	0	0	0.1 ②	0.1	0.1	0.1
5	0	0	0	0	0.2	0.2	0.2	0.2
6	0	0	0	0	0.3	0.3	0.3	0.3
7	0	0	0	0	0.3	0.3	0.3	0.3
8	0	0	0	0	0.4	0.4	0.4	0.4
9	0	0	0	0	0.5	0.5	0.5	0.5
10	0	0	0	0	0.5	0.5	0.5	0.6
11	0	0	0	0	0.6	0.6	0.6	0.6
12	0	0	0	0	0.7	0.7	0.7	0.7
13	0	0	0	0	0.8	0.8	0.8	0.8
14	0	0	0	0	0.8	0.9	0.9	0.9
15	0	0	0	0	0.9	0.9	1	1
16	0	0	0	0	1	1	1	1
17	0	0	0	0	1	1	1	1
18	0	0	0	0	1	1	1	1
19	0	0	0	0	1	1	1	1
20	0	0	0	0	1	1	1	1
21–25	0	0	0	0	2	2	2	2
26–30	0	0	0	0	2 ③	2	2	2
31–35	0	0	0	0	2	2	3	3
36–40	0	0	0	0	3	3	3	3
41–45	0	0	0	0	3	3	3	3
46–50	0	0	0	0	3	4	4	4
51–55	0	0	0	0	4	4	4	4
56–60	0	0	0	0	4	4	4	4
61–65	0	0	0	0	4	4	4	5
66–70	0	0	0	0	5	5	5	④ 5

Note: sustained spread is not expected below BUI 70. Type of fire: surface fire with <10% CFB. ◯ = intensity class.

Table 9.11.
Equilibrium ROS (m/min)
Fire Intensity Class

M-1 boreal mixedwood—leafless, 75% conifer / 25% deciduous

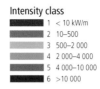

Intensity class
- 1   < 10 kW/m
- 2   10–500
- 3   500–2 000
- 4   2 000–4 000
- 5   4 000–10 000
- 6   >10 000

ISI	0–20	21–30	31–40	41–60	61–80	81–120	121–160	161–200
1	0.2	0.3	0.4	0.4	0.4	0.4	0.5	0.5
2	0.5	0.9	1	1	1	1	1	1
3	0.8	2	2	2	2	2	2	2
4	1	2	3	3	3	3*	3*	4*
5	2	3	4	4	4*	5*	5*	5*
6	2	4	5	5*	6*	6*	6*	6*
7	3	5	6*	7*	7*	7*	8*	8*
8	3	6	7*	8*	8*	9*	9*	9*
9	4	7	8*	9*	10*	10*	11*	12*
10	4	9*	10*	11*	11*	12*	12*	13
11	5	10*	11*	12*	13	13	14	14
12	6	11*	12*	14*	14	15	16	16
13	6	12*	14*	15	16	17	17	18
14	7	13*	15*	16	17	18	19	19
15	7	14*	16	18	19	20	21	21
16	8	15*	18	19	20	21	22	23
17	9	17*	19	21	22	23	24	24
18	9	18	20	22	24	25	25	26
19	10	19	21	24	25	26	27	28
20	10	20	23	25	27	28	29	29
21–25	12	23	26	29	31	32	34	34
26–30	15	29	32	36	38	40	41	42
31–35	18*	34	38	42	44	47	48	49
36–40	20*	38	43	47	50	53	54	55
41–45	22*	42	48	52	56	58	60	61
46–50	24*	46	52	57	60	63	65	67
51–55	26*	49	55	61	65	68	70	71
56–60	27	52	59	64	69	72	74	75
61–65	28	54	62	68	72	75	78	79
66–70	30	57	64	70	75	79	81	82

Constant values: foliar moisture content = 97%; CBH = 6 m. □ = average BUI. Type of fire: Black values = surface with <10% CFB, black values with * = intermittent crown with 10–89% CFB, white values = **continuous crown fire**, — = approximately 50% CFB value. ○ = intensity class.

Table 9.12.
Equilibrium ROS (m/min)
Fire Intensity Class

M-1 boreal mixedwood—leafless, 50% conifer / 50% deciduous

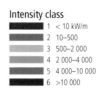

Intensity class
- 1 < 10 kW/m
- 2 10–500
- 3 500–2 000
- 4 2 000–4 000
- 5 4 000–10 000
- 6 >10 000

ISI	\[BUI\] 0–20	21–30	31–40	41–60	61–80	81–120	121–160	161–200
1	0.1	0.2	0.3	0.3	0.3	0.3	0.3	0.3
2	0.3	0.7	0.7	0.8	0.9	0.9	0.9	1
3	0.6	1	②1	1	2	2	2	2
4	0.9	2	2	2	2	2	3	3
5	1	2	3	3	3	3	4*	4*
6	2	3	③4	4	4	4*	5*	5*
7	2	4	4	5	5*	5*	6*	6*
8	2	5	5	6*	6*	7*	7*	7*
9	3	6	6	7*	7*	8*	8*	8*
10	3	6	7*	8*	8*	9*	9*	9*
11	4	7	8*	9*	10*	10*	10*	11*
12	4	8	④9*	10*	11*	11*	12*	12*
13	5	9	10*	11*	12*	13*	13	13
14	5	10*	11*	12*	13*	14	14	14
15	6	11*	12*	13*	14	15	15	16
16	6	12*	13*	14*	15	16	17	17
17	6	12*	⑤14*	15	16	17	18	18
18	7	13*	15*	17	18	18	19	19
19	7	14*	16*	18	19	20	20	21
20	8	15*	17	19	20	21	22	22
21–25	9	18*	20	22	23	24	25	26
26–30	11	22	24	27	29	30	31	31
31–35	13	25	29	32	34	35	36	37
36–40	15	29	33	36	38	40	41	42
41–45	17	32	⑥36	40	42	44	46	47
46–50	18	35	39	43	46	48	50	51
51–55	19	37	42	46	49	52	53	54
56–60	21*	40	45	49	52	55	57	58
61–65	22*	42	47	52	55	58	59	60
66–70	23*	43	49	54	57	60	62	63

Constant values: foliar moisture content = 97%; CBH = 6 m. □ = average BUI. Type of fire: Black values = surface with <10% CFB, black values with * = intermittent crown with 10–89% CFB, white values = **continuous crown fire** , ▬ = approximately 50% CFB value. ◯ = intensity class.

42

Table 9.13.
Equilibrium ROS (m/min)
Fire Intensity Class

Intensity class
1 < 10 kW/m
2 10–500
3 500–2 000
4 2 000–4 000
5 4 000–10 000
6 >10 000

M-1 boreal mixedwood—leafless, 25% conifer / 75% deciduous

				BUI				
ISI	0–20	21–30	31–40	41–60	61–80	81–120	121–160	161–200
1	<0.1 (1)	0.1	0.2	0.2	0.2	0.2	0.2	0.2
2	0.2	0.4	0.5	0.5	0.5	0.6	0.6	0.6
3	0.4 (2)	0.8	0.9	0.9	1	1	1	1
4	0.6	1	1	1	2	2	2	2
5	0.8	2	2	2	2	2	2	2
6	1	2	2	3 (3)	3	3	3	3
7	1	3	3	3	3	4	4	4
8	2	3	3	4	4	4	4*	4*
9	2	4	4	4	5	5*	5*	5*
10	2	4	5	5	6	6*	6*	6*
11	2	5	5	6	6*	7*	7*	7*
12	3	5	6	7 (4)	7*	7*	8*	8*
13	3	6	7	7*	8*	8*	8*	9*
14	3	6	7	8*	9*	9*	9*	9*
15	4	7	8	9*	9*	10*	10*	10*
16	4	8	9	9*	10*	11*	11*	11*
17	4	8	9*	10*	11*	11*	12*	12*
18	5	9	10*	11* (5)	12*	12*	13*	13*
19	5	9	11*	12*	12*	13*	13*	14
20	5	10	11*	12*	13*	14*	14	15
21–25	6	12*	13*	15*	16	16	17	17
26–30	8	15*	16*	18	19	20	21	21
31–35	9	17*	19	21	23	24	25	25
36–40	10	20*	22	24	26	27	28	28
41–45	11	22	25	27 (6)	29	30	31	32
46–50	12	24	27	30	31	33	34	35
51–55	13	26	29	32	34	35	37	37
56–60	14	27	31	34	36	38	39	40
61–65	15	29	32	36	38	40	41	42
66–70	16	30	34	37	40	42	43	44

Constant values: foliar moisture content = 97%; CBH = 6 m. □ = average BUI. Type of fire: Black values = surface with <10% CFB, black values with * = intermittent crown with 10–89% CFB, white values = **continuous crown fire** , ▬ = approximately 50% CFB value. ◯ = intensity class.

Table 9.14.
Equilibrium ROS (m/min)
Fire Intensity Class

M-2 boreal mixedwood—green, 75% conifer / 25% deciduous

Intensity class

Class	kW/m
1	< 10 kW/m
2	10–500
3	500–2 000
4	2 000–4 000
5	4 000–10 000
6	>10 000

	BUI							
ISI	0–20	21– 30	31–40	41–60	61–80	81–120	121–160	161–200
1	0.2	0.3	0.4	0.4	0.4	0.4	0.4	0.5
2	0.5	0.9	(2) 1	1	1	1	1	1
3	0.8	2	2	2	2	2	2	2
4	1	2	(3) 3	3	3	3*	3*	3*
5	2	3	4	4	4*	4*	5*	<u>5*</u>
6	2	4	5	5*	5*	<u>6*</u>	<u>6*</u>	6*
7	3	5	(4) 6	6*	<u>7*</u>	7*	7*	7*
8	3	6	7*	<u>8*</u>	8*	8*	9*	9*
9	4	7	<u>8*</u>	9*	9*	10	10*	10*
10	4	8*	9*	10*	11*	11	12	12
11	5	9*	10*	(5) 12*	12	13	13	14
12	5	<u>10*</u>	12*	13*	14	14	15	15
13	6	11*	13*	14	15	16	16	17
14	7	13*	14*	16	17	17	18	18
15	7	14*	15*	17	18	19	20	20
16	8	15*	17	18	20	20	21	21
17	8	16*	18	20	21	22	23	23
18	9	17*	19	21	22	23	24	25
19	9	18	20	22	24	25	26	26
20	10	19	22	(6) 24	25	26	27	28
21–25	12	22	25	28	29	31	32	32
26–30	14	27	31	34	36	38	39	40
31–35	17	32	36	40	42	44	46	46
36–40	<u>19*</u>	36	41	45	48	50	52	52
41–45	21*	40	45	49	53	55	57	58
46–50	23*	43	49	54	57	60	62	63
51–55	24*	46	52	57	61	64	66	67
56–60	25*	49	55	61	65	68	70	71
61–65	27	51	58	64	68	71	73	75
66–70	28	53	60	66	70	74	76	78

Constant values: foliar moisture content = 97%; CBH = 6 m. □ = average BUI. Type of fire: Black values = surface with <10% CFB, black values with * = intermittent crown with 10–89% CFB, white values = **continuous crown fire**, ___ = approximately 50% CFB value. ◯ = intensity class.

Table 9.15.
Equilibrium ROS (m/min)
Fire Intensity Class

Intensity class

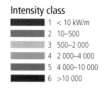

	< 10 kW/m
1	< 10 kW/m
2	10–500
3	500–2 000
4	2 000–4 000
5	4 000–10 000
6	>10 000

M-2 boreal mixedwood—green, 50% conifer / 50% deciduous

	BUI							
ISI	0–20	21–30	31–40	41–60	61–80	81–120	121–160	161–200
1	0.1	0.2	0.2	0.3	0.3	0.3	0.3	0.3
2	0.3	0.6	0.7	0.7	0.8	0.8	0.8	0.9
3	0.6	1	1	1	1	1	2	2
4	0.8	2	2	2	2	2	2	2
5	1	2	2	3	3	3	3	3
6	1	3	3	4	4	4*	4*	4*
7	2	3	4	4	5*	5*	5*	5*
8	2	4	5	5	6*	6*	6*	6*
9	3	5	6	6*	6*	7*	7*	7*
10	3	6	6	7*	7*	8*	8*	8*
11	3	6	7*	8*	8*	9*	9*	9*
12	4	7	8*	9*	9*	10*	10*	10*
13	4	8	9*	10*	10*	11*	11*	11*
14	4	9	10*	11*	11*	12*	12*	12*
15	5	9*	11*	12*	12*	13	13	14
16	5	10*	11*	13*	13*	14	14	15
17	6	11*	12*	13*	14	15	16	16
18	6	12*	13*	14*	15	16	17	17
19	6	12*	14*	15	16	17	18	18
20	7	13*	15*	16	17	18	19	19
21–25	8	15*	17	19	20	21	22	22
26–30	10	19*	21	23	25	26	27	27
31–35	11	22	25	27	29	30	31	32
36–40	13	25	28	31	33	34	35	36
41–45	14	27	31	34	36	38	39	40
46–50	15	30	34	37	39	41	42	43
51–55	17	32	36	39	42	44	45	46
56–60	18	34	38	42	44	47	48	49
61–65	18	35	40	44	47	49	50	51
66–70	19	37	41	46	48	51	52	53

Constant values: foliar moisture content = 97%; CBH = 6 m. □ = average BUI. Type of fire: Black values = surface with <10% CFB, black values with * = intermittent crown with 10–89% CFB, white values = **continuous crown fire** , ▬ = approximately 50% CFB value. ○ = intensity class.

45

Table 9.16.
Equilibrium ROS (m/min)
Fire Intensity Class

M-2 boreal mixedwood—green, 25% conifer / 75% deciduous

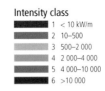

Intensity class
- 1 < 10 kW/m
- 2 10–500
- 3 500–2 000
- 4 2 000–4 000
- 5 4 000–10 000
- 6 >10 000

ISI	0–20	21–30	31–40	41–60	61–80	81–120	121–160	161–200
1	<0.1 ①	0.1	0.1	0.1	0.1	0.2	0.2	0.2
2	0.2	0.3	0.4	0.4	0.4	0.4	② 0.4	0.5
3	0.3	0.6	0.6	0.7	0.7	0.8	0.8	0.8
4	0.4	0.8	1	1	1	1	1	1
5	0.6	1	1	1	2	2	2	2
6	0.8	2	2	2	2	2 ③	2	2
7	1	2	2	2	2	3	3	3
8	1	2	3	3	3	3	3	3
9	1	3	3	3	3	4	4	4
10	2	3	3	4	4	4	4	4*
11	2	3	4	4	4	5	④ 5*	5*
12	2	4	4	5	5	5*	5*	6*
13	2	4	5	5	6	6*	6*	6*
14	2	5	5	6	6*	6*	7*	_7*_
15	3	5	6	6	7*	7*	_7*_	7*
16	3	5	6	7	7*	_8*_	8*	8*
17	3	6	7	7*	8*	8* ⑤	8*	8*
18	3	6	7	8*	_8*_	9*	9*	9*
19	3	7	7	8*	9*	9*	9*	10*
20	4	7	8	9*	9*	10*	10*	10*
21–25	4	8	9*	_10*_	11*	11*	12*	12*
26–30	5	10	_11*_	13*	13*	14*	14	15
31–35	6	12*	13*	15*	16	16	17	17
36–40	7	_13*_	15*	17	18	19	19	19
41–45	8	15*	17*	18	20	21	21	22
46–50	8	16*	18*	20	21	22	⑥ 23	23
51–55	9	17*	20	21	23	24	25	25
56–60	10	18*	21	23	24	25	26	27
61–65	10	19*	22	24	25	27	27	28
66–70	10	20*	23	25	26	28	29	29

Constant values: foliar moisture content = 97%; CBH = 6 m. □ = average BUI. Type of fire: Black values = surface with <10% CFB, black values with * = intermittent crown with 10–89% CFB, white values = **continuous crown fire** , ▬ = approximately 50% CFB value. ◯ = intensity class.

Table 9.17.
Equilibrium ROS (m/min)
Fire Intensity Class

Intensity class

▬	1 < 10 kW/m
▬	2 10–500
▬	3 500–2 000
▬	4 2 000–4 000
▬	5 4 000–10 000
▬	6 >10 000

M-3 dead balsam fir mixedwood— leafless, 30% dead fir

ISI	0–20	21–30	31–40	41–60	61–80	81–120	121–160	161–200
1	0.3	0.5 ②0.6	0.7	0.7	0.8	0.8	0.8	
2	0.7	1	2	2	2	2	2	2
3	1	2 ③3	3	3	3*	3*	4*	
4	2	4	4	4*	5*	5*	5*	5*
5	2	5	5*	6*	6*	6*	7*	7*
6	3	6 ④7*	7*	8*	8*	8*	8*	
7	4	7*	8*	9*	9*	10*	10*	10*
8	4	8*	9*	10*	11*	11*	12	12
9	5	9*	11*	12*	12*	13	13	14
10	5	10* ⑤	12*	13*	14	14	15	15
11	6	12*	13*	14	15	16	17	17
12	7	13*	14*	16	17	18	18	18
13	7	14*	16	17	18	19	20	20
14	8	15*	17	18	20	21	21	22
15	8	16*	18	20	21	22	23	23
16	9	17	19	21	22	23	24	24
17	9	18	20	22	23	25	25	26
18	10	19	21	23	25	26	27	27
19	10	20	22	24	26	27	28	29
20	11	20	23	25	27	28	29	30
21–25	12	23	26	28	30	32	33	33
26–30	14	26 ⑥30	33	35	37	38	38	
31–35	15*	29	33	36	39	41	42	43
36–40	17*	32	36	39	42	44	45	46
41–45	18*	34	38	42	45	47	48	49
46–50	18*	35	40	44	47	49	51	52
51–55	19*	37	42	46	49	51	53	54
56–60	20*	38	43	47	50	53	54	55
61–65	20*	39	44	49	52	54	56	57
66–70	21*	40	45	50	53	55	57	58

BUI (41–60 = average BUI)

Constant values: foliar moisture content = 97%; CBH = 6 m. □ = average BUI. Type of fire: Black values = surface with <10% CFB, black values with * = intermittent crown with 10–89% CFB, white values = **continuous crown fire** , ▬ = approximately 50% CFB value. ◯ = intensity class.

47

Table 9.18.
Equilibrium ROS (m/min)
Fire Intensity Class

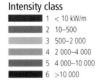

Intensity class
1 < 10 kW/m
2 10–500
3 500–2 000
4 2 000–4 000
5 4 000–10 000
6 >10 000

M-3 dead balsam fir mixedwood— leafless, 60% dead fir

ISI	BUI 0–20	21–30	31–40	41–60	61–80	81–120	121–160	161–200
1	0.5 ②	1	1	1	1	1	1	2
2	1	3 ③	3	3	4*	4*	4*	4*
3	2	5	5	④6*	6*	6*	6*	7*
4	3	6	7*	8*	9*	9*	9*	9*
5	4	9*	10*	⑤11*	11*	12*	12	12
6	6	11*	12*		14	15	15	16
7	7	13*	14*	⑥16	17	18	18	19
8	8	15*	17	18	20	21	21	22
9	9	17	19	21	22	23	24	25
10	10	19	21	24	25	26	27	28
11	11	21	24	26	28	29	30	30
12	12	23	26	28	30	32	33	33
13	13	25	28	31	33	34	35	36
14	14	26	30	33	35	37	38	39
15	15	28	32	35	37	39	40	41
16	16*	30	34	37	40	41	43	44
17	16*	31	36	39	42	44	45	46
18	17*	33	37	41	44	46	47	48
19	18*	35	39	43	46	48	49	50
20	19*	36	41	45	48	50	51	52
21–25	21*	40	45	50	53	55	57	58
26–30	24*	45	51	56	60	63	65	66
31–35	26	50	56	62	66	69	71	72
36–40	28	53	60	66	70	74	76	77
41–45	29	56	63	69	74	78	80	81
46–50	30	58	66	72	77	80	83	85
51–55	31	60	68	74	79	83	86	87
56–60	32	61	69	76	81	85	87	89
61–65	32	62	70	77	82	86	89	91
66–70	33	63	71	78	83	88	90	92

Constant values: foliar moisture content = 97%; CBH = 6 m. □ = average BUI. Type of fire: Black values = surface with <10% CFB, black values with * = intermittent crown with 10–89% CFB, white values = **continuous crown fire** , — = approximately 50% CFB value. ○ = intensity class.

Table 9.19.
Equilibrium ROS (m/min)
Fire Intensity Class

Intensity class
- 1 < 10 kW/m
- 2 10–500
- 3 500–2 000
- 4 2 000–4 000
- 5 4 000–10 000
- 6 >10 000

M-3 dead balsam fir mixedwood—leafless, 100% dead fir

ISI	0–20	21–30	31–40	41–60	61–80	81–120	121–160	161–200
1	0.9	2	2	2	2	2	2	2*
2	2	4	5	5*	6*	6*	6*	6*
3	4	7*	8*	9*	10*	10*	10*	11*
4	5	10*	12*	13*	14	15	15	15
5	7	14*	16	17	18	19	20	20
6	9	17	19	21	23	24	25	25
7	11	21	23	25	27	28	29	30
8	12	24	27	30	31	33	34	35
9	14	27	31	34	36	38	39	39
10	16*	30	34	38	40	42	43	44
11	17*	33	38	41	44	46	48	49
12	19*	36	41	45	48	50	52	53
13	20*	39	44	49	52	54	56	57
14	22*	42	48	52	56	58	60	61
15	23*	45	51	56	59	62	64	65
16	25	47	54	59	63	66	68	69
17	26	50	56	62	66	69	71	73
18	27	52	59	65	69	72	75	76
19	28	54	62	68	72	75	78	79
20	30	57	64	70	75	78	81	82
21–25	33	63	71	78	83	87	89	91
26–30	37	71	80	88	93	98	101	103
31–35	40	77	87	96	102	107	110	112
36–40	43	82	93	102	108	113	117	119
41–45	45	85	97	106	113	118	122	124
46–50	46	88	100	110	117	122	126	128
51–55	47	90	102	112	119	125	129	132
56–60	48	92	104	114	122	127	132	134
61–65	49	93	105	116	123	129	133	136
66–70	49	94	106	117	124	130	135	137

BUI (column header span); 41–60 = average BUI.

Constant values: foliar moisture content = 97%; CBH = 6 m. □ = average BUI. Type of fire: Black values = surface with <10% CFB, black values with * = intermittent crown with 10–89% CFB, white values = **continuous crown fire** . ◯ = intensity class.

Table 9.20.
Equilibrium ROS (m/min)
Fire Intensity Class

M-4 dead balsam fir mixedwood— green, 30% dead fir

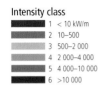

Intensity class
- 1 < 10 kW/m
- 2 10–500
- 3 500–2 000
- 4 2 000–4 000
- 5 4 000–10 000
- 6 >10 000

ISI	0–20	21–30	31–40	41–60	61–80	81–120	121–160	161–200
1	0.1	0.2	0.2	0.3	0.3	② 0.3	0.3	0.3
2	0.3	0.6	0.7	0.7	0.8	③ 0.8	0.8	0.8
3	0.5	1	1	1	1	1	1	1
4	0.8	2	2	2	2	2	2	2
5	1	2	2	3	3	3*	3*	3*
6	1	3	3	3	3*	④ 4*	4*	4*
7	2	3	4	4	4*	4*	5*	5*
8	2	4	4	5*	5*	5*	5*	6*
9	2	4	5	5*	6*	⑤ 6*	6*	6*
10	3	5	6*	6*	7*	7*	7*	7*
11	3	6	6*	7*	7*	8*	8*	8*
12	3	6	7*	8*	8*	9*	9*	9*
13	4	7*	8*	8*	9*	10*	10*	10*
14	4	7*	8*	9*	10*	10*	11*	11*
15	4	8*	9*	10*	11*	11*	11*	12*
16	4	9*	10*	11*	11*	12	12	13
17	5	9*	10*	11*	12	13	13	13
18	5	10*	11*	12*	13	13	14	14
19	5	10*	12*	13*	14	⑥ 14	15	15
20	6	11*	12*	13*	14	15	15	16
21–25	6	12*	14*	15	16	17	18	18
26–30	8	15*	17	18	19	20	21	21
31–35	9	17	19	21	22	23	24	24
36–40	10	18	21	23	24	25	26	27
41–45	10	20	22	25	26	27	28	29
46–50	11	21	24	26	28	29	30	31
51–55	12	22	25	27	29	31	32	32
56–60	12	23	26	28	30	32	33	33
61–65	12	24	27	29	31	33	34	34
66–70	13	24	27	30	32	34	35	35

BUI — 41–60 is the average BUI column.

Constant values: foliar moisture content = 97%; CBH = 6 m. □ = average BUI. Type of fire: Black values = surface with <10% CFB, black values with * = intermittent crown with 10–89% CFB, white values = **continuous crown fire**, ▬ = approximately 50% CFB value. ◯ = intensity class.

Table 9.21.
Equilibrium ROS (m/min)
Fire Intensity Class

M-4 dead balsam fir mixedwood— green, 60% dead fir

Intensity class
1	< 10 kW/m
2	10–500
3	500–2 000
4	2 000–4 000
5	4 000–10 000
6	>10 000

ISI	0–20	21–30	31–40	41–60	61–80	81–120	121–160	161–200
1	0.2	0.4	0.5	0.5	0.5	0.6	0.6	0.6
2	0.6	1	1	1	1	2	2	2
3	1	2	2	2	3	3	3*	3*
4	2	3	3	4	4*	4*	4*	4*
5	2	4	4	5*	5*	6*	6*	6*
6	3	5	6*	6*	7*	7*	7*	7*
7	3	6	7*	8*	8*	9*	9*	9*
8	4	7*	8*	9*	10*	10*	10*	11*
9	4	8*	10*	11*	11*	12*	12	12
10	5	10*	11*	12*	13*	13	14	14
11	6	11*	12*	13*	14	15	15	16
12	6	12*	14*	15	16	17	17	17
13	7	13*	15	16	17	18	19	19
14	7	14*	16	18	19	20	20	21
15	8	15*	17	19	20	21	22	22
16	9	16	19	20	22	23	24	24
17	9	18	20	22	23	24	25	26
18	10	19	21	23	25	26	27	27
19	10	20	22	24	26	27	28	29
20	11	21	23	26	27	29	29	30
21–25	12	23	27	29	31	33	34	34
26–30	14	28	31	35	37	39	40	40
31–35	16*	31	36	39	42	44	45	46
36–40	18*	35	39	43	46	48	50	50
41–45	19*	37	42	46	49	52	53	54
46–50	21*	40	45	49	52	55	57	58
51–55	22*	41	47	51	55	57	59	60
56–60	22*	43	49	53	57	59	61	62
61–65	23*	44	50	55	58	61	63	64
66–70	24*	45	51	56	60	63	65	66

BUI

Constant values: foliar moisture content = 97%; CBH = 6 m. □ = average BUI. Type of fire: Black values = surface with <10% CFB, black values with * = intermittent crown with 10–89% CFB, white values = **continuous crown fire** , ▬ = approximately 50% CFB value. ◯ = intensity class.

51

Table 9.22.
Equilibrium ROS (m/min)
Fire Intensity Class

M-4 dead balsam fir mixedwood— green, 100% dead fir

Intensity class
1 < 10 kW/m
2 10–500
3 500–2 000
4 2 000–4 000
5 4 000–10 000
6 >10 000

ISI	0–20	21–30	31–40	41–60	61–80	81–120	121–160	161–200
1	0.4	0.7 ②0.8	0.8	0.9	0.9	1	1
2	1	2 ③2	2	2	3	3*	3*
3	2	3	4	4	④4*	5*	5*	5*
4	3	5	5*	6*	⑤6*	7*	7*	7*
5	3	7	7*	8*	9*	9*	9*	10*
6	4	8*	9*	10*	11*	12*	12*	12
7	5	10*	11*		13	14	15	15
8	6	12*	14*	15	16	17	17	17
9	7	14*	16	17	⑥18	19	20	20
10	8	16*	18	20	21	22	23	23
11	9	18	20	22	23	25	25	26
12	10	20	22	24	26	27	28	29
13	11	21	24	27	28	30	31	31
14	12	23	26	29	31	32	33	34
15	13	25	28	31	33	35	36	37
16	14	27	30	33	36	37	38	39
17	15*	29	32	36	38	40	41	42
18	16*	30	34	38	40	42	43	44
19	17*	32	36	40	42	44	46	47
20	18*	34	38	42	45	47	48	49
21–25	20*	38	43	48	51	53	55	56
26–30	24*	45	51	56	60	63	65	66
31–35	27	51	58	64	68	71	73	75
36–40	29	56	64	70	75	78	81	82
41–45	32	61	69	75	80	84	87	88
46–50	33	64	73	80	85	89	92	93
51–55	35	67	76	83	89	93	96	98
56–60	36	70	79	86	92	96	99	101
61–65	37	72	81	89	95	99	102	104
66–70	38	73	83	91	97	101	105	107

BUI

Constant values: foliar moisture content = 97%; CBH = 6 m. □ = average BUI. Type of fire: Black values = surface with <10% CFB, black values with * = intermittent crown with 10–89% CFB, white values = **continuous crown fire**, ▬ = approximately 50% CFB value. ◯ = intensity class.

Table 9.23.
Equilibrium ROS (m/min)
Fire Intensity Class

O-1a matted grass

Intensity class
- 1 < 10 kW/m
- 2 10–500
- 3 500–2 000
- 4 2 000–4 000
- 5 4 000–10 000
- 6 >10 000

Degree of curing (%)

ISI	0–20	21–40	41–50	51–60	61–70	71–80	81–90	91–100
1	0 ① <0.1	0.1	0.2	0.4	0.7	1	1	
2	<0.1	0.1	0.3	0.5	1	2	3	3
3	<0.1	0.2	0.5	0.9	2	3	5	6
4	<0.1	0.3	0.7	1	3	5	7	9
5	<0.1	0.4	0.9 ②	2	4	6	9 ③	11
6	<0.1	0.5	1	2	5	8	11	14
7	<0.1	0.6	1	3	6	10	14	18
8	0.1	0.7	2	3	7	12	16	21
9	0.1	0.8	2	4	8	13	19	24
10	0.1	0.9	2	4	9	15	21	27
11	0.1	1	3	5	10	17	24 ④	30
12	0.2	1	3	5	11	19	26	34
13	0.2	1	3	6	13	21	29	37
14	0.2	1	3	6	14	22	31	40
15	0.2	1	4	7	15	24	34	43
16	0.2	1	4	7	16	26	36	46
17	0.2	2	4	8	17	28	39	50
18	0.3	2	4	8	18	30	41 ⑤	53
19	0.3	2	5	9	19	31	43	56
20	0.3	2	5	9	20	33	46	59
21–25	0.3	2	6	11	23	38	53	67
26–30	0.4	3	7	13	27	45	63	81
31–35	0.5	3	8	15	32	52	72	93
36–40	0.5	3	9	16	35	58	81	103
41–45	0.6	4	9	18	38	63	88	113
46–50	0.6	4	10	19	41	68	94	121
51–55	0.6	4	11	20	44	72	100	128
56–60	0.7	4	11	21	46	75	105	134 ⑥
61–65	0.7	4	12	22	48	78	109	140
66–70	0.7	5	12	23	49	81	113	144

Constant values: surface fuel load = 3.5 t/ha. Type of fire: surface. ○ = intensity class.

Table 9.24.
Equilibrium ROS (m/min)
Fire Intensity Class

O-1b standing grass

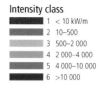

Degree of curing (%)

ISI	0–20	21–40	41–50	51–60	61–70	71–80	81–90	91–100
1	0	<0.1	<0.1	0.1	0.3	0.4	0.6	0.7
2	<0.1	<0.1	0.2	0.4	0.8	1	2	2
3	<0.1	0.1	0.4	0.7	2	3	4	5
4	<0.1	0.2	0.6	1	2	4	6	7
5	<0.1	0.3	0.8	2	3	6	8	10
6	<0.1	0.4	1	2	5	8	10	13
7	<0.1	0.5	1	3	6	10	13	17
8	0.1	0.7	2	3	7	12	16	21
9	0.1	0.8	2	4	8	14	19	25
10	0.1	0.9	2	4	10	16	22	29
11	0.2	1	3	5	11	18	26	33
12	0.2	1	3	6	13	21	29	37
13	0.2	1	3	6	14	23	32	41
14	0.2	1	4	7	15	25	35	45
15	0.2	2	4	8	17	28	39	50
16	0.3	2	4	8	18	30	42	54
17	0.3	2	5	9	20	33	45	58
18	0.3	2	5	10	21	35	49	62
19	0.3	2	6	10	23	37	52	67
20	0.3	2	6	11	24	40	55	71
21–25	0.4	3	7	13	28	47	65	83
26–30	0.5	3	8	16	35	57	80	102
31–35	0.6	4	10	19	41	67	93	120
36–40	0.7	4	11	21	46	76	105	135
41–45	0.7	5	12	23	51	83	116	148
46–50	0.8	5	13	25	55	90	125	160
51–55	0.8	5	14	27	58	95	133	170
56–60	0.9	6	15	28	61	100	140	179
61–65	0.9	6	15	29	64	105	146	187
66–70	1	6	16	30	66	108	150	193

Constant values: surface fuel load = 3.5 t/ha. Type of fire: surface. ◯ = intensity class.

Table 9.25.
Equilibrium ROS (m/min)
Fire Intensity Class

S-1 jack or lodgepole pine slash

Intensity class
1 < 10 kW/m
2 10–500
3 500–2 000
4 2 000–4 000
5 4 000–10 000
6 >10 000

				BUI				
ISI	0–20	21–30	31–40	41–60	61–80	81–120	121–160	161–200
1	0.3	0.6	0.7	0.8	0.9	1	1	1
2	0.7	2	2	2	2	2	2	2
3	1	3	3	3	4	4	4	4
4	2	4	4	5	5	6	6	6
5	2	5	6	6	7	7	8	8
6	3	6	7	8	8	9	9	10
7	3	7	8	9	10	11	11	11
8	4	8	10	11	12	13	13	13
9	4	9	11	13	14	14	15	15
10	5	11	12	14	15	16	17	17
11	5	12	14	16	17	18	19	19
12	6	13	15	17	19	20	21	21
13	6	14	17	19	20	22	23	23
14	7	15	18	20	22	23	24	25
15	7	16	19	22	24	25	26	27
16	8	18	21	23	25	27	28	29
17	8	19	22	25	27	29	30	30
18	9	20	23	26	28	30	31	32
19	9	21	24	28	30	32	33	34
20	9	22	26	29	31	33	35	36
21–25	11	25	29	33	36	38	40	41
26–30	13	30	35	39	42	45	47	48
31–35	15	34	40	45	48	51	54	55
36–40	16	37	44	50	54	57	59	61
41–45	18	41	48	54	58	62	65	66
46–50	19	44	51	58	62	66	69	71
51–55	20	46	54	61	66	70	73	75
56–60	21	48	57	64	69	73	77	78
61–65	22	50	59	66	72	76	80	81
66–70	22	52	61	68	74	79	82	84

□ = average BUI. Type of fire: surface. ◯ = intensity class.

55

Table 9.26.
Equilibrium ROS (m/min)
Fire Intensity Class

S-2 white spruce/balsam slash

Intensity class
1 < 10 kW/m
2 10–500
3 500–2 000
4 2 000–4 000
5 4 000–10 000
6 >10 000

ISI	0–20	21–30	31–40	41–60	61–80	81–120	121–160	161–200
1	<0.1	0.1 ②	0.2	0.2	0.2	0.2	0.2	0.2
2	0.2	0.4	0.5	③0.6	0.6	0.6	0.7	0.7
3	0.4	0.8	1	④1	1	1	1	1
4	0.6	1	1	2	2	2	2	2
5	0.8	2	2	⑤2	3	3	3	3
6	1	2	3	3	3	4	4	4
7	1	3	3	4	4	5	5	5
8	2	4	4	5	5	5	6	6
9	2	4	5	6	6	6	7	7
10	2	5	6	⑥6	7	7	8	8
11	2	6	7	7	8	8	9	9
12	3	6	7	8	9	10	10	10
13	3	7	8	9	10	11	11	11
14	3	8	9	10	11	12	12	12
15	4	9	10	11	12	13	13	13
16	4	9	10	12	13	14	14	14
17	4	10	11	13	14	15	15	16
18	4	10	12	13	15	16	16	17
19	5	11	13	14	16	16	17	18
20	5	11	13	15	16	17	18	19
21–25	6	13	15	17	19	20	21	21
26–30	7	16	19	21	23	24	25	26
31–35	8	18	21	24	26	28	29	29
36–40	9	20	23	26	29	30	32	32
41–45	10	22	25	29	31	33	34	35
46–50	10	23	27	30	33	35	36	37
51–55	10	24	28	32	34	37	38	39
56–60	11	25	29	33	36	38	39	40
61–65	11	26	30	34	37	39	41	42
66–70	11	26	31	35	38	40	42	42

□ = average BUI. Type of fire: surface. ◯ = intensity class.

Table 9.27.
Equilibrium ROS (m/min)
Fire Intensity Class

S-3 coastal cedar/hemlock/ Douglas-fir slash

Intensity class
1 < 10 kW/m
2 10–500
3 500–2 000
4 2 000–4 000
5 4 000–10 000
6 >10 000

					BUI			
ISI	0–20	21–30	31–40	41–60	61–80	81–120	121–160	161–200
1	0	<0.1	<0.1 ②	<0.1	<0.1	<0.1	<0.1	<0.1
2	<0.1	0.1	0.1 ③	0.2	0.2	0.2	0.2	0.2
3	0.2	0.4	0.5 ④	0.5	0.6	0.6	0.6	0.6
4	0.4	0.9	1 ⑤	1	1	1	1	1
5	0.7	2	2	2	2	2	2	3
6	1	2	3	3	4	4	4	4
7	2	4	4	5	5	5	6	6
8	2	5	5 ⑥	6	7	7	8	8
9	3	6	7	8	9	10	10	10
10	3	8	9	10	11	12	13	13
11	4	10	11	13	14	15	15	16
12	5	11	13	15	16	17	18	18
13	6	13	15	17	19	20	21	21
14	6	15	18	20	21	23	24	24
15	7	17	20	22	24	26	27	27
16	8	19	22	25	27	28	29	30
17	9	20	24	27	29	31	32	33
18	10	22	26	29	32	34	35	36
19	10	24	28	31	34	36	38	39
20	11	25	30	33	36	39	40	41
21–25	13	30	35	39	43	45	47	48
26–30	15	36	42	47	51	54	57	58
31–35	17	40	47	53	58	61	64	65
36–40	19	43	51	57	62	66	69	70
41–45	20	45	53	60	65	69	72	73
46–50	20	47	55	62	67	71	74	76
51–55	21	48	56	63	68	73	76	77
56–60	21	48	57	64	69	74	77	79
61–65	21	49	57	65	70	75	78	79
66–70	21	49	58	65	71	75	78	80

□ = average BUI. Type of fire: surface. ○ = intensity class.

Table 10.1.

Spread distance (m)

equilibrium ROS: all fuel types
accelerating ROS: open fuel types
& surface fires in closed fuel types

ROS eq	Elapsed time (min) 5	15	30	45	60	120	180
0.2	1	3 2	6 4	9 7	12 10	24 22	36 34
0.4	2	6 3	12 9	18 15	24 21	48 45	72 69
0.6	3	9 5	18 13	27 22	36 31	72 67	108 103
0.8	4	12 6	24 17	36 29	48 41	96 89	144 137
1	5	15 8	30 22	45 36	60 51	120 111	180 171
2	10	30 16	60 43	90 73	120 103	240 223	360 343
3	15	45 24	90 65	135 109	180 154	360 334	540 514
4	20	60 31	120 86	180 145	240 205	480 445	720 685
5	25	75 39	150 108	225 182	300 257	600 557	900 857
6	30	90 47	180 129	270 218	360 308	720 668	1080 1028
7	35	105 55	210 151	315 254	420 359	840 779	1260 1199
8	40	120 63	240 173	360 291	480 411	960 890	1440 1370
9	45	135 71	270 194	405 327	540 462	1080 1002	1620 1542
10	50	150 79	300 216	450 364	600 513	1200 1113	1800 1713
12	60	180 94	360 259	540 436	720 616	1440 1336	2160 2056
14	70	210 110	420 302	630 509	840 718	1680 1558	2520 2398
16	80	240 126	480 345	720 582	960 821	1920 1781	2880 2741
18	90	270 141	540 388	810 654	1080 924	2160 2003	3240 3083
20	100	300 157	600 432	900 727	1200 1026	2400 2226	3600 3426
25	125	375 196	750 540	1125 909	1500 1283	3000 2783	4500 4283
30	150	450 236	900 647	1350 1091	1800 1539	3600 3339	5400 5139
35	176	525 275	1050 755	1575 1272	2100 1796	4200 3896	6300 5996
40	200	600 314	1200 863	1800 1454	2400 2053	4800 4452	7200 6852
45	225	675 353	1350 971	2025 1636	2700 2309	5400 5009	8100 7709
50	250	750 393	1500 1079	2250 1818	3000 2566	6000 5565	9000 8565
55	275	825 432	1650 1187	2475 1999	3300 2822	6600 6122	9900 9422
60	300	900 471	1800 1295	2700 2181	3600 3079	7200 6678	10800 10278
65	325	975 510	1950 1403	2925 2363	3900 3335	7800 7235	11700 11135
70	350	1050 550	2100 1511	3150 2545	4200 3592	8400 7791	12600 11991
75	375	1125 589	2250 1619	3375 2727	4500 3848	9000 8348	13500 12848
80	400	1200 628	2400 1726	3600 2908	4800 4105	9600 8904	14400 13704
85	425	1275 668	2550 1834	3825 3090	5100 4362	10200 9461	15300 14561
90	450	1350 707	2700 1942	4050 3272	5400 4618	10800 10017	16200 15417
95	475	1425 746	2850 2050	4275 3454	5700 4875	11400 10574	17100 16274
100	500	1500 785	3000 2158	4500 3635	6000 5131	12000 11130	18000 17130

Red values indicate spread distance (m) of an established fire at equilibrium ROS in all fuel types. Values for 5 min define a higher risk zone if a flank fire becomes a head fire. Black values indicate spread distance (m) from a point with accelerating ROS for surface fires (<10% CFB) in closed fuel types and accelerating fires in open fuel types (C-1, D-1, O-1, S-1, S-2, S-3). C-2 and C-7 are also considered open fuel types when the crown closure is less than 50%.

Table 10.2.

Spread distance (m)

accelerating ROS:
crown fires in closed fuel types

ROS eq	Elapsed time (min)											
	15		30		45		60		120		180	
0.2	1	1	3	4	6	7	8	10	20	22	32	34
0.4	2	3	6	8	11	14	17	20	41	44	65	68
0.6	3	4	9	12	17	21	25	30	61	66	97	102
0.8	4	6	12	17	23	28	34	40	81	88	129	136
1	5	7	15	21	28	35	42	50	102	110	162	170
2	9	15	30	42	56	71	84	101	203	221	323	341
3	14	22	46	62	84	106	127	151	305	331	485	511
4	19	30	61	83	113	142	169	202	406	442	646	682
5	24	37	76	104	141	177	211	252	508	552	808	852
6	28	44	91	125	169	213	253	303	609	662	969	1022
7	33	52	106	146	197	248	296	353	711	773	1131	1193
8	38	59	121	167	225	284	338	403	812	883	1292	1363
9	43	67	137	187	253	319	380	454	914	994	1454	1534
10	47	74	152	208	281	355	422	504	1016	1104	1615	1704
12	57	89	182	250	338	426	507	605	1219	1325	1938	2045
14	66	104	212	292	394	497	591	706	1422	1546	2261	2386
16	76	119	243	333	450	568	676	807	1625	1766	2584	2726
18	85	133	273	375	507	639	760	908	1828	1987	2907	3067
20	95	148	303	417	563	710	845	1008	2031	2208	3231	3408
25	118	185	379	521	704	887	1056	1261	2539	2760	4038	4260
30	142	222	455	625	844	1065	1267	1513	3047	3312	4846	5112
35	165	260	531	729	985	1242	1479	1765	3554	3864	5653	5964
40	189	297	607	833	1126	1420	1690	2017	4062	4416	6461	6816
45	213	334	683	937	1266	1597	1901	2269	4570	4968	7269	7668
50	236	371	758	1041	1407	1775	2112	2521	5078	5520	8076	8520
55	260	408	834	1145	1548	1952	2323	2773	5585	6072	8884	9372
60	284	445	910	1250	1689	2130	2535	3025	6093	6624	9692	10224
65	307	482	986	1354	1829	2307	2746	3277	6601	7176	10499	11076
70	331	519	1062	1458	1970	2484	2957	3530	7109	7728	11307	11928
75	355	556	1138	1562	2111	2662	3168	3782	7617	8280	12115	12780
80	378	593	1213	1666	2251	2839	3380	4034	8124	8832	12922	13632
85	402	630	1289	1770	2392	3017	3591	4286	8632	9384	13730	14484
90	426	667	1365	1874	2533	3194	3802	4538	9140	9936	14537	15336
95	449	704	1441	1978	2674	3372	4013	4790	9648	10488	15345	16188
100	473	741	1517	2083	2814	3549	4224	5042	10155	11040	16153	17040

Use this table for initiating crown fires in C2–C-6 and M1–M4 fuel types. Black values indicate spread distances for intermittent crown fires with 50% CFB. Red values indicate spread distances for continuous crown fires with 90% CFB.

Table 11.1.

Fire Area (ha) & Perimeter (m)
C, D, M, S fuel types

Spread distance (m)	Effective wind speed (km/h)										
	0	5	10	15	20	25	30	35	40	45	50
50	0.2	0.2	0.1	<0.1	<0.1	<0.1	<0.1	<0.1	<0.1	<0.1	<0.1
	157	148	133	121	114	110	108	106	105	104	103
100	1	1	0.5	0.4	0.3	0.2	0.2	0.2	0.2	0.1	0.1
	314	297	266	243	229	220	215	212	209	208	206
150	2	2	1	1	1	1	0.5	0.4	0.4	0.3	0.3
	471	445	399	364	343	330	323	317	314	312	310
200	3	3	2	2	1	1	1	1	1	1	1
	628	594	532	486	457	441	430	423	419	415	413
250	5	4	3	2	2	2	1	1	1	1	1
	785	742	665	607	572	551	538	529	523	519	516
300	7	6	5	4	3	2	2	2	1	1	1
	942	891	798	728	686	661	645	635	628	623	619
400	13	11	9	6	5	4	3	3	2	2	2
	1257	1188	1064	971	915	881	860	847	837	831	826
500	20	17	13	10	8	6	5	4	4	4	3
	1571	1485	1330	1214	1144	1102	1075	1058	1047	1038	1032
600	28	25	19	14	11	9	7	6	6	5	5
	1885	1782	1596	1457	1372	1322	1290	1270	1256	1246	1239
800	50	45	34	25	20	16	13	11	10	9	8
	2513	2376	2128	1943	1830	1762	1720	1693	1675	1661	1652
1000	79	70	53	40	31	25	20	18	16	14	13
	3142	2970	2660	2428	2287	2203	2151	2116	2093	2077	2065
1200	113	101	77	57	44	35	29	25	22	20	19
	3770	3564	3192	2914	2745	2644	2581	2540	2512	2492	2478
1500	177	157	120	89	69	55	46	40	35	32	29
	4712	4455	3989	3642	3431	3305	3226	3175	3140	3115	3097
2000	314	279	213	159	122	98	82	71	62	56	52
	6283	5940	5319	4857	4575	4406	4301	4233	4187	4154	4130
2500	491	436	333	248	191	153	128	110	98	88	81
	7854	7425	6649	6071	5718	5508	5376	5291	5233	5192	5162
3000	707	628	479	357	275	221	184	159	140	127	117
	9425	8910	7979	7285	6862	6609	6452	6349	6280	6231	6195
4000	1257	1117	852	635	489	393	328	282	250	226	207
	12566	11880	10638	9713	9149	8812	8602	8466	8373	8307	8260
5000	1963	1746	1331	992	764	613	512	441	390	352	324
	15708	14849	13298	12141	11437	11015	10753	10582	10466	10384	10324
L/B	1.0	1.1	1.5	2.0	2.6	3.3	3.8	4.4	5.0	5.6	6.1
	0	1.4	2.8	4.2	5.6	6.9	8.3	9.7	11.1	12.5	13.9

Effective wind speed (m/s)

Within the table, red values indicate area and black values indicate perimeter length.

Note: C = conifer, D = deciduous, M = mixedwood, S = slash.

Table 11.2.

Fire Area (ha) & Perimeter (m)
O-1 matted and standing grass

Spread distance (m)	Effective wind speed (km/h)										
	0	5	10	15	20	25	30	35	40	45	50
50	0.2	0.1	0.1	0.1	<0.1	<0.1	<0.1	<0.1	<0.1	<0.1	<0.1
	157	117	110	107	106	105	104	104	103	103	103
100	1	0.3	0.2	0.2	0.2	0.2	0.1	0.1	0.1	0.1	0.1
	314	234	220	215	212	210	208	207	206	206	205
150	2	1	1	0.5	0.4	0.4	0.3	0.3	0.3	0.3	0.3
	471	350	330	322	318	315	313	311	310	309	308
200	3	1	1	1	1	1	1	0.5	0.5	0.5	0.5
	628	467	441	430	424	420	417	415	413	411	410
250	5	2	2	1	1	1	1	1	1	1	1
	785	584	551	537	530	525	521	518	516	514	513
300	7	3	2	2	2	1	1	1	1	1	1
	942	701	661	645	635	629	625	622	619	617	615
400	13	5	4	3	3	3	2	2	2	2	2
	1257	935	881	859	847	839	833	829	826	823	821
500	20	8	6	5	4	4	4	3	3	3	3
	1571	1168	1101	1074	1059	1049	1042	1036	1032	1029	1026
600	28	12	9	7	6	6	5	5	5	4	4
	1885	1402	1322	1289	1271	1259	1250	1244	1239	1234	1231
800	50	22	16	13	11	10	9	9	8	8	7
	2513	1869	1762	1719	1694	1678	1667	1658	1651	1646	1641
1000	79	34	25	20	18	16	15	14	13	12	12
	3142	2336	2203	2149	2118	2098	2084	2073	2064	2057	2052
1200	113	49	35	29	26	23	21	20	19	18	17
	3770	2804	2643	2578	2542	2518	2500	2487	2477	2469	2462
1500	177	76	55	46	40	36	33	31	29	27	26
	4712	3505	3304	3223	3177	3147	3126	3109	3096	3086	3077
2000	314	135	98	81	71	64	59	55	52	49	46
	6283	4673	4406	4297	4236	4196	4167	4146	4129	4115	4103
2500	491	211	153	127	111	100	92	86	81	76	73
	7854	5841	5507	5372	5295	5245	5209	5182	5161	5143	5129
3000	707	305	221	183	160	144	133	123	116	110	105
	9425	7009	6609	6446	6354	6294	6251	6219	6193	6172	6155
4000	1257	541	392	325	285	257	236	219	206	195	186
	12566	9346	8812	8595	8472	8392	8335	8291	8257	8229	8206
5000	1963	846	613	508	445	401	368	343	322	305	291
	15708	11682	11015	10743	10590	10490	10419	10364	10322	10287	10258
L/B	1.0	2.3	3.2	3.5	4.4	4.9	5.3	5.7	6.1	6.4	6.8
	0	1.4	2.8	4.2	5.6	6.9	8.3	9.7	11.1	12.5	13.0

Effective wind speed (m/s)

Within the table, red values indicate area and black values indicate perimeter length.

Table 12.1.

Perimeter Growth Rate (m/min)
C, D, M, S fuel types

Effective wind speed (km/h)

ROSeq	0	5	10	15	20	25	30	35	40	45	50
0.2	1	1	1	<1	<1	<1	<1	<1	<1	<1	<1
0.4	1	1	1	1	1	1	1	1	1	1	1
0.6	2	2	2	1	1	1	1	1	1	1	1
0.8	3	2	2	2	2	2	2	2	2	2	2
1	3	3	3	2	2	2	2	2	2	2	2
2	6	6	5	5	5	4	4	4	4	4	4
3	9	9	8	7	7	7	6	6	6	6	6
4	13	12	11	10	9	9	9	8	8	8	8
5	16	15	13	12	11	11	11	11	10	10	10
6	19	18	16	15	14	13	13	13	13	12	12
7	22	21	19	17	16	15	15	15	15	15	14
8	25	24	21	19	18	18	17	17	17	17	17
9	28	27	24	22	21	20	19	19	19	19	19
10	31	30	27	24	23	22	22	21	21	21	21
12	38	36	32	29	27	26	26	25	25	25	25
14	44	42	37	34	32	31	30	30	29	29	29
16	50	48	43	39	37	35	34	34	33	33	33
18	57	53	48	44	41	40	39	38	38	37	37
20	63	59	53	49	46	44	43	42	42	42	41
25	79	74	66	61	57	55	54	53	52	52	52
30	94	89	80	73	69	66	65	63	63	62	62
35	110	104	93	85	80	77	75	74	73	73	72
40	126	119	106	97	91	88	86	85	84	83	83
45	141	134	120	109	103	99	97	95	94	93	93
50	157	148	133	121	114	110	108	106	105	104	103
55	173	163	146	134	126	121	118	116	115	114	114
60	188	178	160	146	137	132	129	127	126	125	124
65	204	193	173	158	149	143	140	138	136	135	134
70	220	208	186	170	160	154	151	148	147	145	145
75	236	223	199	182	172	165	161	159	157	156	155
80	251	238	213	194	183	176	172	169	167	166	165
85	267	252	226	206	194	187	183	180	178	177	176
90	283	267	239	219	206	198	194	190	188	187	186
95	298	282	253	231	217	209	204	201	199	197	196
100	314	297	266	243	229	220	215	212	209	208	206
	0	1.4	2.8	4.2	5.6	6.9	8.3	9.7	11.1	12.5	13.9

Effective wind speed (m/s)

For greater accuracy, use the sum of the head and backfire ROS for ROSeq. Note: C = conifer, D = deciduous, M = mixedwood, S = slash.

Table 12.2.
Perimeter Growth Rate (m/min)
for grass fuel types

	Effective wind speed (km/h)										
ROSeq	0	5	10	15	20	25	30	35	40	45	50
0.2	1	<1	<1	<1	<1	<1	<1	<1	<1	<1	<1
0.4	1	1	1	1	1	1	1	1	1	1	1
0.6	2	1	1	1	1	1	1	1	1	1	1
0.8	3	2	2	2	2	2	2	2	2	2	2
1	3	2	2	2	2	2	2	2	2	2	2
2	6	5	4	4	4	4	4	4	4	4	4
3	9	7	7	6	6	6	6	6	6	6	6
4	13	9	9	9	8	8	8	8	8	8	8
5	16	12	11	11	11	10	10	10	10	10	10
6	19	14	13	13	13	13	13	12	12	12	12
7	22	16	15	15	15	15	15	15	14	14	14
8	25	19	18	17	17	17	17	17	17	16	16
9	28	21	20	19	19	19	19	19	19	19	18
10	31	23	22	21	21	21	21	21	21	21	21
12	38	28	26	26	25	25	25	25	25	25	25
14	44	33	31	30	30	29	29	29	29	29	29
16	50	37	35	34	34	34	33	33	33	33	33
18	57	42	40	39	38	38	38	37	37	37	37
20	63	47	44	43	42	42	42	41	41	41	41
25	79	58	55	54	53	52	52	52	52	51	51
30	94	70	66	64	64	63	63	62	62	62	62
35	110	82	77	75	74	73	73	73	72	72	72
40	126	93	88	86	85	84	83	83	83	82	82
45	141	105	99	97	95	94	94	93	93	93	92
50	157	117	110	107	106	105	104	104	103	103	103
55	173	129	121	118	116	115	115	114	114	113	113
60	188	140	132	129	127	126	125	124	124	123	123
65	204	152	143	140	138	136	135	135	134	134	133
70	220	164	154	150	148	147	146	145	145	144	144
75	236	175	165	161	159	157	156	155	155	154	154
80	251	187	176	172	169	168	167	166	165	165	164
85	267	199	187	183	180	178	177	176	175	175	174
90	283	210	198	193	191	189	188	187	186	185	185
95	298	222	209	204	201	199	198	197	196	195	195
100	314	234	220	215	212	210	208	207	206	206	205
	0	1.4	2.8	4.2	5.6	6.9	8.3	9.7	11.1	12.5	13.9

Effective wind speed (m/s)

For greater accuracy, use the sum of the head and backfire ROS for ROSeq.

Table 13.
Flank Fire Spread Rate (m/min)

	Effective wind speed (km/h)										
ROS$_{tot}$	0	5	10	15	20	25	30	35	40	45	50
0.2	0.1	0.1	0.1	0.1	0.0	0.0	0.0	0.0	0.0	0.0	0.0
0.4	0.2	0.2	0.1	0.1	0.1	0.1	0.1	0.0	0.0	0.0	0.0
0.6	0.3	0.3	0.2	0.2	0.1	0.1	0.1	0.1	0.1	0.1	0.0
0.8	0.4	0.4	0.3	0.2	0.2	0.1	0.1	0.1	0.1	0.1	0.1
1	1	0.5	0.3	0.3	0.2	0.2	0.1	0.1	0.1	0.1	0.1
2	1	1	1	1	0.4	0.3	0.3	0.2	0.2	0.2	0.2
3	2	1	1	1	1	0.5	0.4	0.3	0.3	0.3	0.2
4	2	2	1	1	1	1	1	0.5	0.4	0.4	0.3
5	3	2	2	1	1	1	1	1	1	0.4	0.4
6	3	3	2	2	1	1	1	1	1	1	0.5
7	4	3	2	2	1	1	1	1	1	1	1
8	4	4	3	2	2	1	1	1	1	1	1
9	5	4	3	2	2	1	1	1	1	1	1
10	5	5	3	3	2	2	1	1	1	1	1
12	6	5	4	3	2	2	2	1	1	1	1
14	7	6	5	4	3	2	2	2	1	1	1
16	8	7	5	4	3	2	2	2	2	1	1
18	9	8	6	5	3	3	2	2	2	2	1
20	10	9	7	5	4	3	3	2	2	2	2
25	13	11	8	6	5	4	3	3	3	2	2
30	15	14	10	8	6	5	4	3	3	3	2
35	18	16	12	9	7	5	5	4	4	3	3
40	20	18	13	10	8	6	5	5	4	4	3
45	23	20	15	11	9	7	6	5	5	4	4
50	25	23	17	13	10	8	7	6	5	4	4
55	28	25	18	14	11	8	7	6	6	5	4
60	30	27	20	15	12	9	8	7	6	5	5
65	33	30	22	16	13	10	9	7	7	6	5
70	35	32	23	18	13	11	9	8	7	6	6
75	38	34	25	19	14	11	10	9	8	7	6
80	40	36	27	20	15	12	11	9	8	7	7
85	43	39	28	21	16	13	11	10	9	8	7
90	45	41	30	23	17	14	12	10	9	8	7
95	48	43	32	24	18	14	12	11	10	8	8
100	50	45	34	25	19	15	13	11	10	9	8
L/B	1.0	1.1	1.5	2	2.6	3.2	3.8	4.5	5	5.6	6.1

ROS$_{tot}$: Use head + back fire ROS. O-1 fuel type: Use L/B from Table 11.2 (not effective wind speed). All Other fuel types: Use effective wind speed or L/B from Table 11.1.

References

Alexander, M.E. 1982. Calculating and interpreting forest fire intensities. Can. J. Bot. 60(4): 349–357.

Alexander, M.E. 2010. Surface fire spread potential in trembling aspen during summer in the boreal forest region of Canada For. Chron. 86: 200–212.

Alexander, M.E.; DeGroot, W.J. 1998. Fire behavior in jack pine stands as related to the Canadian Forest Fire Weather Index (FWI) System. Can. For. Serv., North. For. Cent., Edmonton, AB. Poster (with text).

Alexander, M.E.; Lanoville, R.A. 1989. Predicting fire behavior in the black-spruce-lichen woodland fuel type of western and northern Canada. For. Can., North. For. Cent., Edmonton, Alberta and Gov. Northwest Territ. Dep. Renewable Resour., Territ. For. Fire Cent., Fort Smith, N.W.T. Poster (with text).

Alexander, M.E.; Lawson, B.D.; Stocks, B.J.; Van Wagner, C.E. 1984. User guide to the Canadian Forest Fire Behavior Prediction System: rate of spread relationships. Interim ed. Environ. Can., Can. For. Serv., Fire Danger Group, Ottawa, ON.

Alexander, M.E.; Taylor, S.W.; Page, W.G. 2016. Wildland firefighter safety and fire behavior prediction on the fireline. Proc. 13th Int. Wildland Fire Safety Summit and 4th Human Dimensions of Wildland Fire Conference. April 20–24, 2015. Boise, ID. Int. Assoc. Wildland Fire, Missoula, MT.

Andrews, P.L.; Rothermel, R.C. 1982. Charts for interpreting wildland fire behavior characteristics. USDA, For. Serv., Intermt. Res. Stn., Ogden, UT. Res. Pap. INT-RP-131.

Beighley, M. 1995. Beyond the safety zone – creating a margin of safety. Fire Manag. Note 55(4): 21–24.

Byram, G.M. 1959. Forest fire behavior. Pages 90-123 *in* K.P. Davis, ed. Forest fire: control and use. McGraw-Hill, New York.

[CFS]Canadian Forest Service. 1984. Tables for the Canadian Forest Fire Weather Index System. Environ. Can., Can. For. Serv., Ottawa, ON. For. Tech. Rep. 25.

Catchpole, E.A.; Alexander, M.E.; Gill, A.M. 1992. Elliptical-fire perimeter and area-intensity distributions. Can. J. For. Res. 22:968–972.

Cheney, P.; Gould, J.; McCaw, L. 2001. The dead-man zone – neglected area of firefighter safety. Aust. For. 64(1): 45–50.

Cole, F.V.; Alexander, M.E. 1995. Head fire intensity class graph for FBP System fuel type C-2 (Boreal Spruce). Alaska Dep. Nat Resour., Div. For., Fairbanks, Alaska and Can. For. Serv., North. For. Cent., Edmonton, AB. Poster (with text).

Cova, T.J.; Dennison, P.E.; Kim, T.H.; Moritz, M.A. 2005. Setting wildfire evacuation trigger points using fire spread modeling and GIS. Trans. GIS. 9: 603–617.

De Groot, W.J. 1993. Examples of fuel types in the Canadian Forest Fire Behavior Prediction (FBP) System. For. Can., Northwest Reg., North For. Cent., Edmonton, AB. Poster (with text).

Forestry Canada Fire Danger Group. 1992. Development and structure of the Canadian Forest Fire Behavior Prediction System. For. Can., Ottawa, ON. Inf. Rep. ST-X-3.

Hirsch, K. G. 1996. Canadian Forest Fire Behavior Prediction (FBP) System: users's guide. Nat. Resour. Can., Can. For. Serv., Northwest Reg., North. For. Cent., Edmonton, AB. Spec. Rep. 7.

Kidnie, S.M.; Wotton, B.M.; Droog, W.N. 2010. Field guide for predicting fire behavior in Ontario's tallgrass prairie. Nat. Resour. Can., and Ont. Minist. Nat. Resour., Sault Ste. Marie, ON.

Lawson, B.D.; O.B. Armitage; Hoskins, W.D. 1996. Diurnal variation in the Fire Fuel Moisture Code: tables and source code. Can. For. Serv., Pac. For. Cent., Victoria, BC and B.C. Minist. For., Res. Branch, Victoria, BC. Canada-British Columbia Partnership Agreement on Forest Resource Development: FRDA II. FRDA Rep. 245.

Lawson, B.D.; Armitage, O.B. 2008. Weather guide for the Canadian Forest Fire Danger Rating System. 2008. Nat. Resour. Can., Can. For. Serv., North. For. Cent., Edmonton, AB.

List, R.J. 1951. Smithsonian meterological tables. 6th rev. ed. Smithsonian Inst. Press, Washington, DC.

Merrill, D.F. ; Alexander, M.E., Eds. 1987. Glossary of forest fire management terms. 4th ed. Natl. Res. Counc. Can., Can. Comm. For. Fire Manag., Ottawa, ON. Publ. NRCC 265216.

Pearce, H.G.; Anderson, S.A.J. 2008. A manual for predicting fire behavior in New Zealand fuels. SCION, Rural Fire Res. Group, Christchurch, NZ.

Rothermel, R. C. 1991. Predicting behavior and size of crown fires in the Northern Rocky Mountains. U.S. Dep. Agric., For. Serv., Ogden, UT. Res. Pap. INT-438.

Tymstra, C.; Bryce, R.W.; Wotton, B.M.; Taylor S.W.; Armitage, O.B. 2010. Development and structure of Prometheus – the Canadian Wildand Fire Growth Model. Nat. Resour. Can., Can. For. Serv., North. For. Cent., Edmonton, AB. Inf. Rep. NOR-X-417.

Van Wagner, C. E. 1969. A simple fire-growth model. For. Chron. 4: 103–104.

Wotton, B.M.; Alexander, M.E.; Taylor, S.W. 2009. Updates and revisions to the 1992 Canadian Forest Fire Behavior Prediction System. Nat. Resour. Can., Can. For. Serv., Gt. Lakes For. Cent., Sault Ste. Marie, ON. Inf. Rep. GLC-X-10.

Appendix 1.
Abbreviations

BISI	Backfire initial spread index
BROS	Backfire rate of spread (m/min)
BUI	Buildup Index
CBH	Crown base fire
CC	Continuous crown fire
CFB	Crown fraction burned
DC	Drought code
DMC	Duff moisture code
FFMC	Fine Fuel Moisture Code
IC	Intermittent crown fire
ISI	Initial Spread Index
L/B	Length/breadth ratio
PC	Percent conifer
PDF	Percent dead fir
PD	Percent deciduous
ROS	Head fire rate of spread (m/min)
ROSeq	Equilibrium rate of spread (m/min)
S	Surface fire
WD	Wind direction (degrees)
WSE	Effective wind speed (km/h)
WSV	Net effective wind speed (km/h)

Appendix 2.

Glossary

aspect—The direction a slope is facing (see slope azimuth).

Buildup Index (BUI)—A numerical rating of the total amount of fuel available for combustion that combines Duff Moisture Code (DMC) and Drought Code (DC).

convection column—The definable plume of hot gases, smoke, firebrands, and other combustion by-products produced by and rising above a fire.

crown base height (CBH)—The height, above ground, where the live crown of coniferous trees begins. This value is constant for each FBP System fuel type except C-6.

crown fraction burned (CFB)—The crown fraction burned is the proportion of tree crowns involved in the fire in a given area. The following descriptive classes are recognized in the FBP System:

Crown fraction burned	Type of fire
< 10%	surface fire
10–89%	intermittent crown fire
> 90%	continuous crown fire

degree of curing—The proportion of cured and/or dead plant material in a grassland fuel complex.

Duff Moisture Code (DMC)—A numerical rating of the average moisture content of loosely compacted organic layers of moderate depth. This code indicates the fuel consumption in moderate duff layers and medium-sized woody material.

Drought Code (DC)— A numerical rating of the average moisture content of deep, compact organic layers. This code indicates seasonal drought effects on forest fuels, and the amount of smoldering in deep duff layers and large logs.

effective wind speed (WSE)—The sum of the vectors of the 10-m open wind speed and the slope equivalent wind speed.

elliptical fire growth model—Theory: a free-burning point source fire will spread with an elliptical shape when fuels are uniform and continuous, topography is homogenous, the wind direction is constant (but non-zero), and the fire is unaffected by suppression activities.

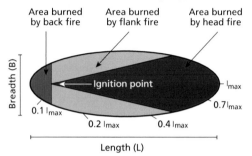

The length of the ellipse is the sum of the head fire spread distance and the backfire spread distance. Fire shape or length-to-breadth (L/B) ratio is determined by the wind speed, which becomes narrower with increasing effective wind speed. Fire area and perimeter length are calculated from the total spread distance and the L/B ratio. Fire spread rate, intensity and flame length are at a maximum at the head of the fire (1_{max}) and decrease around the perimeter to a minimum at the back of the fire.

Fine Fuel Moisture Code (FFMC)— A numerical rating of the moisture content of litter and other cured fine fuels. This code indicates the relative ease of ignition and flammability of fine fuel. The standard daily FFMC is calculated from noon weather observations but represents fine fuel moisture at 1600 LST. The **diurnal FFMC** is an estimate of the FFMC at a particular hour, based on the typical daily variation in temperature and relative humidity in the absence of rain. The **hourly FFMC** is calculated from hourly weather observations. Fire behavior varies with the diurnal cycle in fuel moisture and in wind speed. Fine fuel moisture content is usually at a maximum just before dawn and reaches a minimum in the late afternoon; FFMC has an opposite trend. The peak in FFMC extends longer in the day at high latitudes in summer; this is reflected in the hourly but not diurnal FFMC. The daily FFMC trend may also vary with slope and aspect, peaking earlier on east slopes and later on west slopes, especially in open fuel types, however this is not presently accounted for by either the diurnal or hourly FFMC.

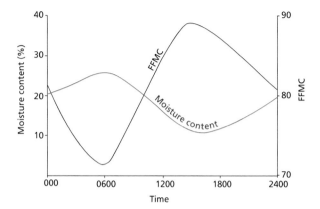

fire intensity—The rate of heat energy release per unit time per unit length of fire front. Frontal fire intensity is a major determinant of certain fire effects and difficulty of control. Numerically, it is equal to the product of the net heat of combustion, the quantity of fuel consumed in the flaming front, and the linear rate of spread.

fire perimeter—The entire outer edge or boundary of a fire. See elliptical fire growth model.

flame angle—The angle formed between the flame at the **fire front** and the ground surface, expressed in degrees.

flame depth —The width of the zone within which continuous flaming occurs behind the edge of a **fire front**.

flame height—The average maximum vertical extension of flames at the **fire front**; occasional flashes that rise above the general level of flames are not considered.

flame length—The length of flames measured along their axis at the **fire front**; the distance between the **flame height** tip and the midpoint of the **flame depth** at the ground surface. **Flame length** is an approximate indicator of **frontal fire intensity**. Below: cross-sectional view of the flame front illustrating the flame length (Lf), flame height (Hf), flame angle (Af), and depth of the flame zone (Df).

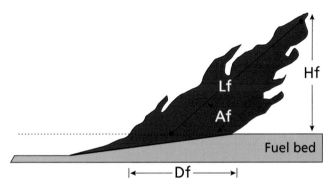

foliar moisture content—The percent moisture content by weight of live conifer needles that are at least 1-year old.

forest floor—The surface layer of fire-carrying fuel including needles and leaf litter, dead grass litter, lichen and feather moss.

Initial Spread Index (ISI)—A numerical rating of the expected rate of fire spread. It combines the effects of wind and Fine Fuel Moisture Code on rate of spread, but excludes the influence of the variable quantities of fuel. ISI can have different values in the FWI and FBP Systems. In the FBP System the effect of slope is accounted for as a slope equivalent wind speed, which is added to the ambient wind to obtain the effective wind speed; this is then used to derive ISI.

inversion—The atmospheric condition in which temperature within a vertical layer of air increases with altitude, resulting in a very stable atmosphere until the inversion lifts or breaks. This is contrary to the usual situation in which temperature decreases with height. Temperature inversions at the earth's surface are a common occurrence in the early morning hours during the fire season and dampen fire behavior.

ladder fuel—Fuels that provide vertical continuity between the surface and crown fuels in a forest stand, thus contributing to the ease of torching and crowning (e.g., tall shrubs, small-sized trees, bark flakes, tree lichens).

length-to-breadth ratio (L/B)—See elliptical fire growth model.

low-level jet wind—A particular type of wind condition aloft, evident in the vertical wind profile, in which there is a zone of increasing wind speed near the earth's surface, and a zone of decreasing velocity above a point of maximum wind speed. Working values for the "jet point" height and wind speed maximum are roughly 500 metres (m) and 30+ kilometres per hour (km/h), respectively. Low-level jets may interact with the convection column on large fires, increasing convective circulation and fire intensity, or may mix down and increase the surface wind speed in the afternoon period.

organic layer—The accumulated partially to fully decomposed organic matter at the soil surface. It corresponds to the fermentation (F) and humus (H) layers in forests and/or the O (peat) layer in wetlands.

rate of spread (ROS)—The speed at which a fire extends its horizontal dimensions, expressed in terms of distance per unit of time. Generally thought of in terms of a fire's forward movement, or head fire rate of spread, but also applicable to backfire and flank fire ROS. As a rule of thumb, flank fire ROS is approximately ½ head fire ROS midway between the head and back of the fire.

relative humidity (RH)—The ratio, expressed as a percentage, of the amount of water vapor or moisture in the air to the maximum amount of moisture that the air could hold at the same dry-bulb temperature and atmospheric pressure (RH can vary from 0 to 100%).

safety margin (SM)—The safety margin is the cushion of time in excess of the time needed by fire fighters to get to a safety zone before the fire gets to them:

safety margin = escape time – fire arrival time

where escape time is the distance along the escape route to the safety zone divided by the fire fighter's rate of travel, and the fire arrival time is the distance from the fire to the safety zone divided by the rate of spread in that direction.

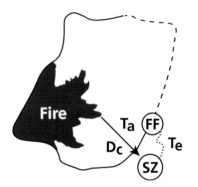

FF firefighter's present location
SZ safety zone
T_a fire arrival time
T_e escape route travel time
D_c critical distance
— constructed fireline
--- planned fireline
······· escape route

If the safety margin and escape time are predetermined, the critical distance when the safety margin will be reached can be estimated as:

critical distance = rate of spread X (escape time + safety margin).

A similar principle is used to determine evacuation trigger points. These procedures should be used with caution as it is difficult to accurately estimate or predict many of the values in the field.

Example: Assuming a fire front 300 m away is spreading in a C-2 fuel type and ISI 8, BUI 80. ROS is 10 m/min and the escape time is 24 min.

safety margin = 30 min – 24 min = + 6 min. The fire fighters will be in the safety zone for 6 minutes before the fire reaches the safety zone.

critical distance = 10 m/min x (24 min + 6 min) = 300 m. The critical distance between the fire location and the fire fighters is 300 m for the example conditions.

situational awareness—The perception of environmental conditions with respect to time or space, the comprehension of their meaning, and the projection of changing conditions over time or space. Situational awareness comprises the first two phases of the observe-orient-decide-act cycle.

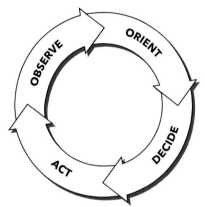

slope azimuth—The uphill slope direction, 180° opposite the slope aspect (if aspect ≤ 180°, slope azimuth = aspect + 180° ; if aspect > 180°, slope azimuth = aspect − 180°.

slope equivalent wind speed—An approach used in the FBP System whereby the effect of slope on fire spread with zero wind is given a value in units of wind speed.

spread azimuth—The direction in which the fire is spreading, determined by combining the wind and slope azimuths.

stand composition—The proportion of each tree species in a stand expressed as a percentage of the total; in the fire management sense, as a percentage of the crown biomass.

stand structure—The horizontal and vertical distribution of components of a forest stand including the crown layers and stems of trees, shrubs, herbaceous understory, snags, and down woody debris.

surface fuel—All combustible materials lying above the duff layer between the ground and ladder fuels that are responsible for propagating surface fires (e.g., litter, herbaceous vegetation, low and medium shrubs, tree seedlings, stumps, downed-dead roundwood).

type of fire—See crown fraction burned.

wind azimuth—The direction the wind is blowing, that is 180° opposite the wind direction (the direction the wind is coming from). If WD ≤ 180° then wind azimuth = wind direction + 180°; if WD > 180° then wind azimuth = WD − 180°.

Appendix 3.
Selected unit conversion factors

If the SI units are	Multiply by	To obtain	Inverse factor
Hectares (ha)	2.4711	Acres (ac)	0.40469
Kilometres per hour (km/h)	0.62137	Miles per hour (mi/h)	1.6093
Kilometres per hour (km/h)	0.2778	Metres per second (m/s)	3.6
Kilowatts per metre (kW/m)	0.28909	Btu per second per foot (Btu/s/ft)	3.4592
Metres (m)	0.049709	Chains (ch)	20.117
Metres (m)	3.2808	Feet (ft)	0.3048
Metres per minute (m/min)	3.2808	Feet per minute (ft/min)	0.3048
Metres per minute (m/min)	2.9826	Chains per hour (ch/h)	0.33528
Metres per minute (m/min)	60.0	Metres per hour (m/h)	0.016667
Metres per minute (m/min)	0.06	Kilometres per hour (km/h)	16.667
Tonnes per hectare (t/ha)	0.44609	Tons per acre (T/ac)	2.2417

Note: all factors are given to five significant digits. If fewer digits are given, the value is exact. To convert Imperial or old metric unit values to the International System of Units (SI), multiply by the inverse factor given in the right-hand column. A "Btu" is a British thermal unit.

Appendix 4.
FBP System fuel type photographs

C-1 Spruce–Lichen Woodland

Open black spruce stands with lichen understory on well-drained upland sites.

Crown	Open park-like black spruce stands occurring as widely spaced individuals and in dense clumps
	Minor amounts of jack pine and white birch in the overstory
	Tree heights vary but live and dead branches extend to the forest floor
	Branch layering is extensive
Surface	Woody surface fuel accumulation is very light and scattered
	Shrub cover is very sparse
Forest floor	Ground surface is fully exposed the sun
	Nearly continuous cover of reindeer lichen averaging 3–4 cm depth above mineral soil

C-2 Boreal Spruce

Upland and lowland black spruce, white and Engelmann spruce stands. Does not include spruce-sphagnum bogs.

Crown
Pure, moderately well-stocked black spruce stands, typically with flaky bark on the lower bole

Tree crowns extend to or near the ground

Dead branches draped with bearded lichens (*Usnea* spp.)

Surface
Low shrubs present, often Labrador tea

Low to moderate volumes of downed woody debris

Forest floor
Nearly continuous cover of feather mosses and/or lichens (e.g., reindeer lichen)

Sphagnum mosses may occur in small amounts but do not affect fire spread

Compact organic layer commonly exceeds 20–30 cm depth

C-3 Mature Jack or Lodgepole Pine

Fully stocked mature jack and lodgepole pine stands.

Crown	Pure fully stocked jack pine or lodgepole pine stands (1000–2000 stems/ha: intertree spacing ~2–3 m)
	Complete crown closure
	Base of live crown is well above the ground
Surface	Sparse conifer understory may be present
	Light and scattered dead surface fuels
Forest floor	Feather mosses over a moderately deep (10 cm) compacted organic layer

C-4 Immature Jack or Lodgepole Pine

Densely stocked immature jack and lodgepole pine stands with nearly continuous horizontal and vertical fuel layers

Crown
: Pure, dense jack or lodgepole pine stands (10 000–30 000 stems/ha: intertree spacing ≤ 1 m)

Large quantity of standing dead stems (from natural thinning)

Surface
: Large quantity of dead down woody fuel

Needle litter and needles suspended within a low (*Vaccinium* sp.) shrub layer

Surface fuel loadings are greater than in the C-3 fuel type

Forest floor
: Organic layers are shallower (<10 cm) and less compact than C-3

C-5 Red and White Pine

Mature stands of red pine and eastern white pine. This fuel type has been applied to tall, mature, closed canopy stands of coastal Douglas-fir and western redcedar/western hemlock.

Crown	Mature stands of red pine and eastern white pine
	Small components of white spruce and old white birch or aspen
	Moderately dense understory, usually red maple or balsam fir
Surface	Shrub layer, usually beaked hazel, in moderate proportions
	Ground surface cover of herbs and pine litter
Forest floor	Organic layer is usually 5–10 cm deep

C-6 Conifer Plantation

Red pine plantations. This fuel type is applicable to all conifer plantations with closed crown canopy and no understory shrub layer.

Crown	Pure conifer plantations, fully stocked with closed crowns
	Rate of spread and crown fire relationships vary with crown base height
Surface	No understory or shrub layer present
	Ground surface cover of needle litter
Forest floor	Duff layer is up to 10 cm deep

C-7 Ponderosa Pine/Douglas-fir

Mixed stands of uneven aged ponderosa pine and Douglas-fir.

Crown
Uneven-aged stands of ponderosa pine and Douglas-fir

Western larch and lodgepole pine may be significant stand components

Stands are open with occasional clumpy thickets of multi-aged Douglas-fir and/or larch as a discontinuous understory

Canopy closure is less than 50% overall although thickets are often closed and often dense

Surface
Woody surface fuel accumulations are light and scattered

Ground surface is dominated by perennial grasses, herbs, and scattered shrubs

Needle litter in tree thickets

Forest floor
Duff layers are shallow (< 3 cm) in thickets and absent in openings

D-1 Leafless Aspen

Pure semi-mature trembling aspen stands in the leafless stage.

Crown   Pure semi-mature trembling aspen stands before budbreak in the spring or following leaf fall and curing of understory vegetation in the autumn

Conifer understory is notably absent

Surface   Well-developed medium to tall shrub layer is typically present

Dead and down roundwood fuels are light

Fire spread is mainly in deciduous leaf litter and cured herbaceous material that is directly exposed to wind and solar radiation

Forest floor   Duff (F and H horizons) seldom contribute to the available fuel due to high moisture content in the spring

D-2 Green Aspen

Pure semi-mature trembling aspen stands in the green stage.

Crown	Pure semi-mature trembling aspen stands after leaf flush in the spring–summer
	Conifer understory is notably absent
Surface	Well-developed medium to tall shrub layer is typically present
	Dead and down roundwood fuels are light
	Fire spread is mainly in deciduous leaf litter
Forest floor	Duff (F and H horizons) up to 5 cm deep

M-1 Boreal Mixedwood—Leafless

Mixed stands of boreal conifers and deciduous species on upland sites in the leafless stage. Photo shows a stand with approximately 75% coniferous and 25% deciduous component.

Crown	Mixed stands of variable proportions including
	Conifers – black spruce, white spruce, balsam fir, subalpine fir
	Deciduous – trembling aspen and white birch
	Individual species may be absent on particular sites
	Stand structure (height and size) is also variable; conifer crowns extend nearly to ground
	Rate of spread is weighted according to the proportion of conifers and deciduous species
	M-1 leafless stage occurs in spring and fall
Surface	Feather mosses, conifer needles, and deciduous leaves
Forest floor	Moderately compact organic layer, up to 8–10 cm deep

M-2 Boreal Mixedwood—Green

Mixed stands of boreal conifers and deciduous species in the green stage.
Photo shows a stand with approximately 75% coniferous and 25%
deciduous component.

Crown	Mixed stands of variable proportions including
	Conifers – black spruce, white spruce, balsam fir, subalpine fir
	Deciduous – trembling aspen and white birch
	Individual species may be absent on particular sites
	Stand structure (height and size) is also variable; conifer crowns extend nearly to ground
	Rate of spread is weighted according to the proportion of conifers and deciduous species
	M-2 green stage occurs in summer
Surface	Moderate shrub and continuous herb layers
	Low to moderate dead, down woody fuels; scattered to moderate conifer understory
Forest floor	Moderately compact organic layer, up to 8–10 cm deep

M-3 Dead Balsam Fir/Mixedwood—Leafless

Mixed stands of dead balsam fir and boreal mixedwood species in the leafless stage. This photo shows a stand with approximately 60% dead balsam fir and 40% live mixedwood components. Note that percent dead fir is the percentage of the stand composed of dead fir, not the percentage of fir that is dead.

Crown	Mixedwood stands of spruce, pine, birch, and balsam fir with abundant budworm-killed balsam fir in the understory
	Peeling bark, top breakage, windthrow, and development of arboreal lichens (old man's beard) peaking 5–8 years after tree mortality
	M-3 leafless stage occurs in the spring and fall. After tree mortality, spring fires are very vigorous with continuous crowning and downwind spotting
Surface	Volume of down woody material is initially low, increasing substantially with stand breakup following tree mortality
	Feather mosses, conifer needles, and deciduous leaves
Forest floor	Moderately compact organic layer 8–10 cm deep

M-4 Dead Balsam Fir/Mixedwood—Green

Mixed stands of dead balsam fir and boreal mixedwood species in the green stage. This photo shows a stand with approximately 60% dead balsam fir and 40% live mixedwood components. Note that percent dead fir is the percentage of the stand composed of dead fir, not the percentage of fir that is dead.

Crown   Mixedwood stands of spruce, pine, birch, and balsam fir with abundant budworm-killed balsam fir in the understory

Peeling bark, top breakage, windthrow, and development of arboreal lichens (old man's beard) peaking 5–8 years after tree mortality

M-4 green stage occurs in summer

Surface   Volume of down woody material is initially low, increasing substantially with stand breakup following mortality

Summer fires are initially hampered by the proliferation of green understory vegetation resulting from the opening of a stand; however, fires will spread through the fuel complex when sufficient surface fuel accumulates (usually after 4–5 years). Fire behavior is greatest 5–8 years following tree mortality, decreasing as surface woody fuels decompose and understory vegetation develops

Feather mosses, conifer needles, and deciduous leaves

Forest floor   Organic layer moderately compacted 8–10 cm deep

O-1 Grass

Matted and standing grass. This photo shows well-cured standing grass.

Surface   Continuous grass cover

Occasional trees and/or shrubs that do not appreciably affect the fire spread rate, but may contribute to spotting and breaching of fire guards

Matted grass condition (0-1a) common in spring after snow melt

Standing dead grass condition (0-1b) common in late summer to early fall

The proportion of cured or dead material in grasslands has a pronounced effect on fire spread and must be estimated with care

Organic Layer   Absent or very shallow

S-1 Jack or Lodgepole Pine Slash

Jack or lodgepole pine slash, one- to two-seasons old.

Surface   Continuous slash resulting from clearcut logging of mature jack or lodgepole pine stands

Slash is typically one- or two-seasons old, retaining up to 50% of the needles, particularly on branches closest to the ground

No post-logging treatment has been applied and slash is continuous

Tops and branches left on site result in moderate fuel load and depth

Continuous feather moss with discontinuous fallen needle litter

Forest floor   Organic layers are moderately deep and fairly compact

S-2 White Spruce/Balsam Slash

White or Engelmann spruce and balsam or subalpine fir slash, one- to two-seasons old.

Surface	Slash resulting from clearcut logging of mature to overmature stands of white spruce and subalpine fir or balsam fir
	Slash is typically one- to two-seasons old, retaining from 10% to 50% of the foliage on the branches
	No post-logging treatment has been applied
	Fuel continuity may be broken by skid trails
	Tops left on site, branches broken off, resulting in moderate slash loads and depths
	Quantities of large and rotten woody debris may be significant
	Feather moss with additional needle litter from slash
Forest floor	Organic layers are moderately deep and compact

S-3 Coastal Cedar/Hemlock/Douglas-fir Slash

Western red cedar, western hemlock, and Douglas-fir slash, one-season old in coastal BC. This fuel type is also applicable to interior wet belt cedar-hemlock slash, and may be used for blow down in the absence of other information.

Surface	Slash resulting from clearcut logging of mature to overmature mixed western red cedar, western hemlock and Douglas-fir stands
	Slash is typically one-season old with cured foliage
	Cedar foliage present; hemlock and Douglas-fir will have dropped up to 50% of their needles
	Slash fuels continuous and uncompacted, 0.5–2.0 m deep
	Very large loadings of large, broken and rotten debris may be present
	Minor to moderate shrub and herbaceous understory may be present
	Feather moss or old compact needle litter and recent needle litter fallen from the slash
Forest floor	Organic layers are moderately deep to deep and compact

Appendix 5.
Summary of FBP System fuel type characteristics

	Forest floor and organic layer	Surface and ladder fuels	Stand structure and composition
C-1	**Spruce–Lichen Woodland** Continuous reindeer lichen; organic layer absent or shallow, uncompacted.	Very sparse herb shrub cover and down woody fuels; tree crowns extend to ground.	Open black spruce with dense clumps; associated species include jack pine, white birch; well-drained upland sites.
C-2	**Boreal Spruce** Continuous feather moss and/or Cladonia lichen; deep, compacted organic layer.	Continuous shrub (e.g., Labrador tea); low to moderate downed woody fuels; tree crowns extend nearly to ground; arboreal lichens, flaky bark.	Moderately well-stocked black spruce stands on both upland and lowland sites; Sphagnum bogs excluded.
C-3	**Mature Jack or Lodgepole Pine** Continuous feather moss; moderately deep, compacted organic layer.	Sparse conifer understory may be present; sparse down woody fuels; tree crowns separated from ground.	Fully stocked jack or lodgepole pine stands; mature.
C-4	**Immature Jack or Lodgepole Pine** Continuous needle litter; moderately shallow organic layer.	Moderate shrub/herb cover; continuous vertical crown fuel continuity; heavy standing dead and down woody fuel.	Dense jack or lodgepole pine stands; immature.
C-5	**Red and White Pine** Continuous needle litter; moderately shallow organic layer.	Moderate herb and shrub (e.g., hazel); moderately dense understory (e.g., red maple, balsam fir); tree crowns separated from ground.	Moderately well-stocked red and white pine stands; mature; associated species: white spruce, white birch, and aspen.
C-6	**Conifer Plantation** Continuous needle litter; moderately shallow organic layer.	Absent herb/shrub cover, absent understory; tree crowns separated from ground.	Fully stocked conifer plantations; complete crown closure regardless of mean stand height; mean stand crown base height controls ROS and crowning.

	Forest floor and organic layer	Surface and ladder fuels	Stand structure and composition
C-7	**Ponderosa Pine/Douglas-fir** Continuous needle litter; absent to shallow organic layer.	Discontinuous grasses/herbs except in conifer thickets, where absent; light woody fuels; tree crowns separated from ground except in thickets.	Open ponderosa pine and Douglas-fir stands; mature uneven-aged; associated species of western larch, lodgepole pine; understory conifer thickets.
D-1/2	**Aspen** Continuous leaf litter; shallow, uncompacted organic layer.	Moderate medium to tall shrubs and herb layers; absent conifer understory; sparse, dead, down woody fuels.	Moderately well-stocked trembling aspen stands; semi-mature; leafless (i.e., spring, fall, or following disease or insect outbreak); or green in summer after budburst.
M-1/2	**Boreal Mixedwood** Continuous leaf litter in deciduous portions of stands; discontinuous feather moss and needle litter in conifer portions of stands; organic layers shallow, uncompacted to moderately compacted.	Moderate shrub and continuous herb layers; low to moderate dead, down woody fuels; conifer crowns extend nearly to ground; scattered to moderate conifer understory.	Moderately well-stocked mixed stand of boreal conifers (e.g., black/white spruce, balsam/subalpine fir) and deciduous species (e.g., trembling aspen, white birch). Fuel types are differentiated by season and percent conifer/deciduous species composition.
M-3/4	**Dead Balsam Fir/Mixedwood** Continuous leaf litter in deciduous portions of stands; discontinuous feather moss, needle litter and deciduous leaves in mixed portions of stands; organic layers moderately compacted, 8–10 cm.	Dense continuous herbaceous cover after greenup; down woody fuels low initially, but becoming heavy several years after balsam mortality; ladder fuels dominated by dead balsam understory.	Moderately well-stocked mixed stand of spruce, pine and birch with dead balsam fir, often as an understory. Fuel types differentiated by season and age since balsam mortality.

		Forest floor and organic layer	Surface and ladder fuels	Stand structure and composition
O-1	**Grass**	Organic layer absent to shallow and moderately compacted.	Continuous grass fuel; standard loading is 0.35 kg/m^2, but other loading can be accommodated; percent cured or dead must be estimated. Continuous dead grass litter. Sparse or scattered shrubs and downed woody fuel. Conditions for both early spring matted grass and late summer; standing cured grass are included.	Scattered trees, if present, do not appreciably affect fire spread rate but may contribute to spotting and breaching of fire guards.
S-1	**Jack or Lodgepole Pine Slash**	Continuous feather moss; discontinuous needle litter; moderately deep, compacted organic layer.	Continuous slash, moderate loading and depth; high foliage retention; absent to sparse shrub and herb cover.	Slash from clearcut logging; mature jack or lodgepole pine stands.
S-2	**White Spruce/Balsam Slash**	Continuous feather moss and needle litter; moderately deep, compacted organic layer.	Continuous to discontinuous slash (due to skidder trails); moderate foliage retention; moderate loading and depth; moderate shrub and herb cover.	Slash from clearcut logging; mature or overmature white spruce, subalpine fir or balsam fir stands.
S-3	**Coastal Cedar/Hemlock/ Douglas-fir Slash**	Continuous feather moss or compacted old needle litter below fresh needle litter from slash; moderately deep to deep, compacted organic layer.	Continuous slash, high foliage retention (cedar), moderate for other species; heavy loading, deep slash; sparse to moderate shrub and herb cover.	Slash from clearcut logging; mature to overmature cedar, hemlock, or Douglas-fir stands.

Flame length (m) Fire intensity class for surface fires

Intensity class		Flame length (m)
1	< 10 kW/m	<0.2
2	10–500	0.2–1.4
3	500–2 000	1.4–2.6
4	2 000–4 000	2.6–3.5
5	4 000–10 000	3.5–5.3
6	>10 000	>5.3

Flame height (m)	Flame angle (degrees) 90	75	60	45	30	15
<0.2	<0.2	0.2	0.2	0.3	0.4	0.8
0.4	0.4	0.4	0.5	0.6	0.8	1.5
0.6	0.6	0.6	0.7	0.8	1.2	2.3
0.8	0.8	0.8	0.9	1.1	1.6	3.1
1.0	1.0	1.0	1.2	1.4	2.0	3.9
1.2	1.2	1.2	1.4	1.7	2.4	4.6
1.4	1.4	1.4	1.6	2.0	2.8	5.4
1.6	1.6	1.7	1.8	2.3	3.2	6.2
1.8	1.8	1.9	2.1	2.5	3.6	7.0
2.0	2.0	2.1	2.3	2.8	4.0	7.7
2.5	2.5	2.6	2.9	3.5	5.0	9.7
3.0	3.0	3.1	3.5	4.2	6.0	11.6
3.5	3.5	3.6	4.0	4.9	7.0	13.5
4.0	4.0	4.1	4.6	5.7	8.0	15.5
4.5	4.5	4.7	5.2	6.4	9.0	17.4
5.0	5.0	5.2	5.8	7.1	10.0	19.3
5.5	5.5	5.7	6.4	7.8	11.0	21.3
6.0	6.0	6.2	6.9	8.5	12.0	23.2
6.5	6.5	6.7	7.5	9.2	13.0	25.1
7.0	7.0	7.2	8.1	9.9	14.0	27.0
7.5	7.5	7.8	8.7	10.6	15.0	29.0
8.0	8.0	8.3	9.2	11.3	16.0	30.9
8.5	8.5	8.8	9.8	12.0	17.0	32.8
9.0	9.0	9.3	10.4	12.7	18.0	34.8
9.5	9.5	9.8	11.0	13.4	19.0	36.7
10.0	10.0	10.4	11.5	14.1	20.0	38.6

Use to estimate flame length and fire intensity class from the flame height and angle. ◯ = intensity class.

Appendix 7.
Fire behavior descriptions

Intensity class (kW/m)		Fire behavior description
1	<10	Smoldering ground or creeping surface fire Little visible flame Firebrands and active fires tend to be self-extinguishing except with high DC and/or BUI
2	10–500	Low vigor surface fire. In stands with low CBH, some foliage of individual trees consumed
3	500–2 000	Moderately vigorous surface fire with both low and high flames Ladder fuels (lichen and bark flakes) consumed Isolated torching in stands with low CBH or ladder fuels
4	2 000–4 000	Highly vigorous surface fire with moderate to high flames Passive crowning (isolated to abundant torching) increasing with amount of ladder fuels and low CBH
5	4 000–10 000	Extremely vigorous surface fire or active crown fire with abundant torching and continuous crowning in dense stands Flames extend from the forest floor to above the tree canopy Short to medium range spotting likely
6	>10 000	Blow-up or conflagration type fire runs Continuous crowning in forested fuel types Great walls of flame Towering convection columns Medium to long-range spotting Fire whirls

Fire behavior prediction worksheet example

1	Fire no./Name	Sample		Date	29.04.05	Time	1700
2	Prediction date & time DD/MM/YY		30.04.05	From	1300	To	1400
3	Prediction point & ignition type		1 PI	2 PI	3 LS	4 LS	5 PI
	Fuel type						
4	Fuel type identifier		C-4	M-1	D-1	S-2	O-1a
5	Modifiers (CBH; PC; PDF)			75:25			
	Fine Fuel Moisture Code						
	Yesterday's FFMC						
6	Standard daily FFMC		93	94	88	94	94
7	Diurnal or hourly FFMC		91	92	85	92	92
8	Ground slope (%)		0	15	10	50 (15)	20 (L)
9	Aspect		N.A.[a]	N	E	S (E)	W (L)
10	Adjusted FFMC		91	92	85	95	92
	Wind and Initial Spread Index						
11	Slope equiv. wind speed (km/h)		0	5	3	26	8
12	10-m wind speed (km/h)		23	15	-3	5	-15
13	Effective wind speed (km/h)		23	20	0	31	-7
14	ISI – head/back		16/2	16/2	2/2	41/2	8/4
	Rate of Spread and Intensity						
	Yesterday's DMC						
	Today's DMC						
	Yesterday's DC						
	Today's DC						
15	BUI or Degree of Curing (%)		100	25	60	100	95
16	Equilibrium ROS (m/min) – head		27	15	0.2	33	21
17	– back		2	0.9	0.2	0.6	9
18	Fire intensity class – head/back		6/3	5/2	2/2	6/4	4/3
19	Type of fire – head/back		CC/S	IC/S	S/S	S/S	S/S
20	CFB (%) – head /back		90/10	80/10	N.A.	N.A.	N.A.
	Fire size						
21	Elapsed time (min)		60	60	60	60	60
22	Head fire spread distance (m)		1362	757	12	1980	1077
23	Backfire spread distance (m)		103	46	12	36	482
24	Total spread distance (m)		1465	803	24	2016	1539
25	Elliptical fire area (ha)		62	20	N.A.	N.A.	72
26	Elliptical fire perimeter (m)		3350	1830	N.A.	N.A.	3514
27	L/B ratio		3.0	2.6	N.A.	N.A.	2.7
28	Perimeter growth rate (m/min)		61	41	N.A.	N.A.	4.8

[a]N.A. = not applicable.

Fire behavior prediction worksheet

1	Fire no./Name		Date		Time		
2	Prediction date & time DD/MM/YY		From		To		
3	Prediction point & ignition type						
Fuel type							
4	Fuel type identifier						
5	Modifiers (CBH; PC; PDF)						
Fine Fuel Moisture Code							
	Yesterday's FFMC						
6	Standard daily FFMC						
7	Diurnal or hourly FFMC						
8	Ground slope (%)						
9	Aspect						
10	Adjusted FFMC						
Wind and Initial Spread Index							
11	Slope equiv. wind speed (km/h)						
12	10-m wind speed (km/h)						
13	Effective wind speed (km/h)						
14	ISI – head/back						
Rate of Spread and Intensity							
	Yesterday's DMC						
	Today's DMC						
	Yesterday's DC						
	Today's DC						
15	BUI or Degree of Curing (%)						
16	Equilibrium ROS (m/min) – head						
17	– back						
18	Fire intensity class – head/back						
19	Type of fire – head/back						
20	CFB (%) – head /back						
Fire size							
21	Elapsed time (min)						
22	Head fire spread distance (m)						
23	Backfire spread distance (m)						
24	Total spread distance (m)						
25	Elliptical fire area (ha)						
26	Elliptical fire perimeter (m)						
27	L/B ratio						
28	Perimeter growth rate (m/min)						

Probabilities of Sustained Ignition in Lodgepole Pine, Interior Douglas-fir, and White Spruce – Subalpine Fir Forest Types

B.D. Lawson and G.N. Dalrymple

Fire Management Network
Canadian Forest Service
Northern Forestry Centre
1998

Introduction

Tables of sustained ignition probabilities are presented for four benchmark forest fuel types: dry- and moist-site lodgepole pine, white spruce–subalpine fir, and interior Douglas-fir. The tables were derived from logistic regression equations that were fit to hundreds of small-scale match-fire and campfire observations made near Prince George and 100 Mile House, B.C. by Canadian Forest Service fire researchers. The field research studies were carried out in the 1950s and 1960s as part of the early development of the Canadian Forest Fire Danger Rating System, but the original data have been re-analyzed using current versions of components of the Canadian Forest Fire Weather Index System as predictors of ignition probability.

Derivation of Equations

A complete scientific and technical description of the development of the lodgepole pine and white spruce–subalpine fir equations is given in Lawson et al. (1994a), and is illustrated in a wall poster (Lawson et al. 1994b) as graphs with color photos of the forest types. A computer application (Lawson, Armitage, and Dalrymple, 1996) links these same equations with a new diurnal Fine Fuel Moisture Code program (Lawson, Armitage, and Hoskins 1996).

The interior Douglas-fir equation for probability of sustained ignition used to produce Table 4 was derived from analysis of a data set collected, during the years 1957–1959 near 100 Mile House, B.C. by Dr. P.M. Paul, retired research scientist, Canadian Forest Service. The test-fire program there resulted in publication of the Cariboo, British Columbia, Forest Fire Danger Tables in 1965, while the study sites themselves were described in detail in annual progress reports (Paul, 1957, 1958, 1959), which were used for decisions about data selection for the present analysis and for the site photograph.

The data for site BC-801 were obtained from a national data base on test fires conducted in Canada between 1931 and 1961 as part of a Canadian Forest Service program of regional forest fire danger table development, which preceded the present national fire danger rating system. The test fire data base, consisting of some 20 000 fires, has been edited, merged with current version FWI System component values, and assembled in modem computer formats, as described by Lynham and Martell (1989), and Lynham (1992).

Analysis of 100 Mile House data

Data analysis followed the same steps described by Lawson et al. (1994a). Two suitable study sites were chosen for characterization as interior Douglas-fir from the nine available; specifically, forest type "D-1, Firgrass/moss" (site code 80104) and "D-4, Fir–grass/needle" (site 80107) (photo of D-4 included with Table 4). Site descriptions are included here as they were presented in Paul (1957):

> **Site D1** – Douglas-fir in varying age classes. Large mature fir predominates, but there is an abundance of older immature (81–120 years) with well-established young growth in understory . . .
> Intermingled with the grass, which is ubiquitous except under the trees, are patches of moss . . .

> **Site D4** – is located on the southwest slope of the ridge at a slightly lower elevation than D1, which occurs on a bench near the summit of the hill. It is well stocked, with the fir trees attaining a very good size, although smaller than the mature and overmature trees of D1. Young reproduction, up to 4 feet, is extensive, and there is a fairly complete ground cover of grass, although it is continuously broken by patches of fir needles.

Paul (1957, 1958, and 1959) reported progression of grass curing through the test-fire seasons. Curing started in late May 1957 as 75–80%, decreasing to 35–45% by mid-July, and increasing again during September to about 70% by late September. However, in 1958 (a drier summer) grass was 75–85% cured by late August. Grass curing could not be analyzed as an independent variable in the model, but an assumption could be made that test fires were usually conducted when grass was 50% or more cured.

Before the test-fire data were analyzed for the two sites, diurnally adjusted FFMC and ISI were calculated for each test fire using the diurnal FFMC program from Lawson, Armitage, and Hoskins (1996) and the noon 10-m open wind speed recorded at the headquarters weather station. A dependent binary variable, ignition, was defined as yes if the burned area for match fires was more than 0.30 ft^2, and no if less than or equal to 0.30 ft^2. This area criterion was chosen arbitrarily after inspecting the data and comparing them with results from a separate analysis on vigor class (on file, Pacific Forestry Centre). While the burned-area threshold for successful ignition was larger than the 0.05 ft² used by Lawson et al. (1994a) for lodgepole pine match fires, the interior Douglas-fir forest types were faster spreading, being grass dominated.

Logistic regression was used to develop a preliminary model of ignition as predicted from diurnally adjusted FFMC, DMC, DC, ISI, and BUI. A total of 391 match fires (no camp fires) were available from sites D-1 and D-4 combined, with all D-1 fires recording an area greater than 0, while 80% of D-4 fires burned an area greater than 0.

A logistic regression equation that included ISI and BUI produced the statistically best results from the two independent variables (adding DMC increased the r² and total correct statistics slightly).

The equation for probability of sustained ignition in interior Douglas-fir is plotted in three dimensions in Figure 1, and Table 1 summarizes the observed and expected outcomes by decile and overall goodness-of-fit.

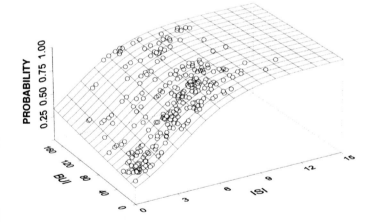

Figure 1. Probability of sustained ignition vs. Initial Spread Index (ISI) and Buildup Index (BUI) for interior Douglas-fir forest type. Plotted points represent test fires from sites D-1 and D-4, 100 Mile House, B.C.

Table 1. Logistic regression statistics: observed (Obs.) and expected (Exp.) outcomes by decile, overall goodness-of-fit, and success of prediction by the model, interior Douglas-fir forest type

P Upper limit	P Avg.	Total fires	Ignition Obs.	Ignition Exp	No ignition Obs.	No ignition Exp
0.307	0.262	38	13	10.0	25	28.0
0.414	0.350	39	17	13.7	22	25.3
0.589	0.509	40	16	20.4	24	19.6
0.657	0.616	39	18	24.0	21	15.0
0.743	0.704	39	27	27.5	12	11.5
0.790	0.766	38	29	29.1	9	8.9
0.843	0.820	40	36	32.8	4	7.2
0.883	0.860	39	36	33.5	3	5.5
0.923	0.902	38	35	34.3	3	3.7
1.000	0.947	41	37	38.8	4	2.2
		391				

Note: χ^2 (8df) = 13.1; P value = 0.106; and R^2 = 0.189.

Success of prediction by the model

	Ignition predicted	No ignition predicted	Total
Ignition	197.6	66.4	264
No ignition	66.4	60.6	127
Predicted total	264.0	127.0	391
Proportion correct	0.749	0.477	N/A[a]
Total correct	0.661	N/A	N/A
Success index	0.073	0.153	N/A

[a]N/A = not applicable.

4

The following four equations are used to calculate the ignition probabilities presented in Tables 2-5:

Dry lodgepole pine
$$P = 1/(1 + \exp(2.107 - 0.727 \times ISI)).\ldots\ldots\ldots\ldots \text{(Table 2)}$$

Moist lodgepole pine
$$P = 1/(1 + \exp(2.146 - 0.009 \times BUI - 0.349 \times ISI))..\ \text{(Table 3)}$$

White spruce–subalpine fir
$$P = 1/(1 + \exp(2.766 - 0.005 \times DC - 0.396 \times ISI))\ ..\ \text{(Table 4)}$$

Interior Douglas-fir
$$P = 1/(1 + \exp(1.563 - 0.005 \times BUI - 0.478 \times ISI))..\ \text{(Table 5)}$$

Probability of sustained ignition is not a standard output from the Canadian Forest Fire Behavior Prediction (FBP) System (Forestry Canada Fire Danger Group 1992), but it is included here as a supplement for the convenience of users. Potential uses include gauging the likelihood of fire starts or fire arrivals for preparedness planning of initial attack crew positioning and readiness, and judging spot-fire ignition potential for wildfires or prescribed fires. The ignition probabilities described here relate to the probability of an ignition from small fire brands, such as matches or campfires persisting with flaming combustion (for at least 2 minutes for match fires and 15 minutes for campfires) and reaching the threshold condition for fire spread and growth. Camp fires were necessary for sustained ignition in the white spruce–subalpine fir forest type. Matches and camp fires were both potentially successful fire brands in the other forest types, depending on burning conditions.

Use of the Tables

The following steps form the procedure for using the Ignition Probability Tables (Table and Appendix numbers refer to Taylor et al. 1996).

1. Select the most appropriate forest type and/or variant from the four presented.

   Detailed forest type descriptions and a photograph accompany each ignition probability table. Ignition probability differences among variants within a single FBP System fuel type should be expected. The dry and moist lodgepole pine forest types can be considered variants of C-3 (Mature Jack or Lodgepole Pine). The interior Douglas-fir forest type should be considered a variant of C-7 (Ponderosa pine–Douglas-fir). The white spruce–subalpine fir forest type can be considered a variant of C-2 (Boreal spruce), as shown by De Groot (1993).

2. Determine the most representative Fine Fuel Moisture Code (FFMC), 10-m open wind speed, and Initial Spread Index (ISI) for the time of interest.

   a) The hourly FFMC (Van Wagner 1977) should be used, if available. Alternatively, the diurnal FFMC (Tables A.1 – A.2, Appendix 5) may be used for any desired time of the day. The daily standard FFMC may be used to represent mid-afternoon (16:00 hr LST) conditions.

   b) Wind speed at 10 m in the open should be measured, forecasted, or estimated from the Beaufort Scale (Appendix 4) for the time of interest.

   c) For the Dry Lodgepole Pine forest type, enter Table 1 with FFMC and 10-m wind speed from 2a and 2b, above.

      For other forest types and/or variants, determine the ISI using Table 3, with FFMC and 10-m open wind speed from 2a and 2b, above.

3. Determine the Buildup Index (BUI) or Drought Code (DC) from the nearest representative weather station, as required for a given forest type.

4. From the ignition probability table for the forest type of interest, find the probability-of-ignition value (%) at the intersection of the two required input values. Probability-of-ignition classes (low, medium, and high) are differentially shaded in the body of each table.

Dry lodgepole pine forest-type characteristics

- Pure, moderately dense stands of mature lodgepole pine;
- scattered pine understory;
- sandy well-drained soils, low-productivity sites dominated by feathermoss and *Cladina* (lichen); B.C. biogeoclimatic classification SBSmk1/03 is typical;
- forest floor shallow (3-5 cm), low bulk density; and
- fire-carrying fuel—lichen, bearberry (*Arctostaphylos uva-ursi*), feathermoss (*Pleurozium schreberi* and *Dicranum* spp.), needle litter.

Table 2
Dry lodgepole pine forest type-probability of sustained ignition (%) and ignition class

10-m open wind speed (km/h)

FFMC	0	5	10	15	20	25	30	35	40	45	≥50
70	16	18	21	24	30	38	49	63	79	91	97
72	17	19	21	26	32	40	53	68	82	93	98
74	17	19	23	27	34	44	57	73	87	95	99
76	18	21	24	30	38	49	64	79	91	97	99
78	19	23	27	34	44	58	73	87	95	99	100
80	22	26	32	41	54	69	84	94	98	100	100
82	26	32	40	52	68	82	93	98	100	100	100
84	32	40	53	68	83	93	98	100	100	100	100
86	41	54	69	84	94	98	100	100	100	100	100
88	56	71	85	95	99	100	100	100	100	100	100
90	73	87	95	99	100	100	100	100	100	100	100
92	88	96	99	100	100	100	100	100	100	100	100
94	97	99	100	100	100	100	100	100	100	100	100
96	99	100	100	100	100	100	100	100	100	100	100
≥98	100	100	100	100	100	100	100	100	100	100	100

	Ignition class	Probability %
	Low	0–49
	Medium	50–75
	High	76–100

Moist lodgepole pine forest-type characteristics

- Mixed, well-stocked stands of mature lodgepole pine–Douglas-fir;
- significant understory of lodgepole pine, trembling aspen, white spruce, and subalpine fir;
- moderately well-drained soils typically derived from unstratified glacial deposits, medium site productivity, *Cornus canadensis* (bunchberry), scattered *Cladina* (lichen), feather mosses, and *Vaccinium membranaceum* (black huckleberry); B.C. biogeoclimatic classification SBSmk1/05 is typical;
- forest floor shallow (3–5 cm), low bulk density; and
- fire-carrying fuels—feather moss, needle litter.

Table 3
Moist lodgepole pine forest type—probability of sustained igniton (%) and ignition class

Buildup Index

ISI	0–20	21–30	31–40	41–60	61–80	81–120	121–160	161–200
0.5	13	15	16	18	21	26	33	41
1	15	17	19	21	24	29	37	46
1.5	18	20	21	24	27	33	41	50
2	20	23	24	27	31	37	45	54
2.5	23	26	28	31	34	41	50	59
3	27	29	31	34	38	45	54	63
4	34	37	39	43	47	54	62	70
5	42	46	48	51	56	62	70	77
6	51	54	57	60	64	70	77	83
7	60	63	65	68	72	77	83	87
8	68	70	72	75	78	82	87	91
9	75	77	79	81	84	87	91	93
10	81	83	84	86	88	90	93	95
11	86	87	88	90	91	93	95	96
12	89	91	91	92	94	95	96	97
13	92	93	94	94	95	96	97	98
14	94	95	95	96	97	97	98	99
15	96	96	97	97	98	98	99	99
18	99	99	99	99	99	99	100	100

Ignition class	Probability %
Low	0–49
Medium	50–75
High	76–100

White spruce–subalpine fir forest-type characteristics

- Mixed, moderately dense stands of mature white (and/or Engelmann) spruce–subalpine fir, with scattered mature birch and Douglas-fir;
- significant understory of subalpine fir;
- soils range from loamy sands to sticky, poorly drained clays on moderately well-drained to imperfectly drained sites; herb and shrub layers dominated by *Aralia nudicaulis* (sarsaparilla), *Gymnocarpium dryopteris* (oak fern), *Vaccinium* (blueberry), and *Cornus* (bunchberry) spp., *Oplopanax horridum* (devil's club) on wettest sites; mosses patchy; B.C. biogeoclimatic classification SBSmk1/07 (drier) and SBSmkl/08 (wetter) are typical;
- forest floors moderate to deep (6–12 cm), moderate to high bulk density; and
* fire-carrying fuels—litter, feather moss.

Table 4
White spruce–subalpine fir forest type—probability of sustained ignition (%) and ignition class

Drought Code

ISI	0–100	101–200	201–300	301–400	401–500	501–600	601–700	701–800
0.5	9	14	21	31	42	55	66	77
1	11	17	25	35	47	59	71	80
1.5	13	19	28	40	52	64	75	83
2	15	23	33	44	57	68	78	86
2.5	18	26	37	49	62	73	81	88
3	21	30	42	54	66	76	84	90
4	28	39	52	64	74	83	89	93
5	37	49	61	72	81	88	92	95
6	47	59	70	80	87	91	95	97
7	56	68	78	85	91	94	96	98
8	66	76	84	90	93	96	97	98
9	74	82	89	93	95	97	98	99
10	81	87	92	95	97	98	99	99
11	86	91	94	97	98	99	99	100
12	90	94	96	98	99	99	99	100
13	93	96	97	98	99	99	100	100
14	95	97	98	99	99	100	100	100
15	97	98	99	99	100	100	100	100
18	99	99	100	100	100	100	100	100

Ignition class	Probability %
Low	0–49
Medium	50–75
High	76–100

Interior Douglas-fir forest-type characteristics

- Well-stocked stands of interior Douglas-fir of varying age classes, including large, mature overstory trees with a significant understory;
- forest floor shallow (less than 3 cm); and
- fire-carrying fuels—grass covers the ground except under fir thickets, where needle litter and patchy moss predominates. Grass is assumed to be greater than 50% cured (dead), although the degree of curing is not a variable used in the ignition probability model.

Table 5
Interior Douglas-fir—probability of sustained ignition (%) and ignition class

Buildup Index (BUI)

ISI	0–20	21–30	31–40	41–60	61–80	81–120	121–160	161–200
0.5	22	23	24	25	27	30	35	40
1	26	28	29	30	32	36	40	45
1.5	31	33	34	36	38	41	46	51
2	36	38	39	41	44	47	52	57
2.5	42	44	45	47	50	53	58	63
3	48	50	51	53	56	59	64	68
4	60	62	63	65	67	70	74	78
5	71	72	73	75	76	79	82	85
6	79	81	81	83	84	86	88	90
7	86	87	88	88	89	91	92	94
8	91	92	92	92	93	94	95	96
9	94	95	95	95	96	96	97	97
10	96	97	97	97	97	98	98	98
11	98	98	98	98	98	99	99	99
12	99	99	99	99	99	99	99	99
13	99	99	99	99	99	99	100	100
14	99	99	100	100	100	100	100	100
15	100	100	100	100	100	100	100	100

Ignition class	Probability %
Low	0–49
Medium	50–75
High	76–100

References

De Groot, W.J. 1993. Examples of fuel types in the Canadian Forest Fire Behavior Prediction (FBP) System. For. Can., Northwest Reg., North For. Cent., Edmonton, Alberta Poster (with text).

Forestry Canada Fire Danger Group. 1992. Development and structure of the Canadian Forest Fire Behavior Prediction System. For. Can., Sci. Sustainable Dev. Dir., Ottawa, Ontario, Inf. Rep. ST-X-3.

Lawson, B.D.; Armitage, O.B.; Dalrymple, G.N. 1994a. Ignition probabilities for simulated people-caused fires in B.C.'s lodgepole pine and white spruce–subalpine fir forests. Pages 493-505. in Proc. 12th Conf. Fire For. Meteorol; October 26-29, 1993, Jekyll Island, Georgia. Soc. Am. For., Bethesda, Maryland, SAF Publ. 94-02.

Lawson, B.D.; Armitage, O.B.; Dalrymple, G.N. 1994b. Ignition probabilities for lodgepole pine and spruce–subalpine fir forests. Nat. Resour. Can., Can. For. Serv., Pac. For. Cent., Victoria, British Columbia and B.C. Minist. For., Res. Branch, Victoria, British Columbia. Canada–British Columbia Partnership Agreement on Forest Resource Development: FRDA II, FRDA Poster (with text).

Lawson, B.D.; Armitage, O.B.; Dalrymple, G.N. 1996. Wildfire Ignition Probability Predictor (WIPP). R&D Update. Nat. Resour. Can., Can. For. Serv., Pac. Yukon Reg., Pac. For. Cent., Victoria, British Columbia.

Lawson, B.D.; Armitage, O.B.; Hoskins, W.D. 1996. Diurnal variation in the Fine Fuel Moisture Code: tables and computer source code. Nat. Resour. Can., Can. For. Serv., Pac. For. Cent., Victoria, British Columbia and B.C. Minist. For., Res. Branch, Victoria, British Columbia. Canada–British Columbia Partnership Agreement on Forest Resource Development, FRDA II, FRDA Rep. 245.

Lynham, T.J.; Martell, D.L. 1989. Preliminary report on a national database of experimental fires in Canada. Pages 41–44 in: Proc. Natl. Workshop on forest fire occurrence prediction, May 3–4, 1989, Winnipeg, Manitoba.

Lynham, T.J. 1992. Summary of data manipulations to Forestry Canada's 20 000 test-fire data base for addition of Canadian FWI values. For. Can., Ont. Reg., Great Lakes For. Cent., GLFC, Sault Ste. Marie, Ontario. File rep.

Paul, P.M. 1957, 1958, 1959. Progress report on forest fire research, 100 Mile House, B.C., Can. For. Serv., For. Fire Res. Inst., Ottawa, Ontario, File 668.

Taylor, S.W.; Pike, R.G.; Alexander, M.E. 1997. Field guide to the Canadian Forest Fire Behavior Prediction (FBP) System. Nat. Resour. Can., Can. For. Serv., North. For. Cent., Edmonton, Alberta. Spec. Rep. 11.

Van Wagner, C.E. 1977. A method of computing fine fuel moisture content throughout the diurnal cycle. Fish. Environ. Can., Can. For. Serv., Petawawa For. Exp. Stn., Chalk River, Ontario. Inf. Rep. PS-X-69.

Acknowledgments

The authors thank Marty Alexander, Canadian Forest Service (CFS), Edmonton, for suggesting presentation of the logistic regression equations for probability of ignition as tables to complement the existing poster/graph and computer applications, and for reviewing the manuscript. Steve Taylor, CFS, Victoria, and Judi Beck, British Columbia Forest Service Protection Program, also provided thorough and helpful reviews. Tim Lynham, CFS, Sault Ste. Marie, provided the computerized national data set from which the 100 Mile House data about interior Douglas-fir was extracted for analysis. University of Victoria co-op students Jim Andrews and Robin Pike analyzed the data.

Lawson, B.C.; Dalrymple, G.N. 1998. Probabilities of sustained ignition in lodgepole pine, interior Douglas-fir and white spruce–subalpine fir forest types. Nat. Resour. Can., Can. For. Serv., North. For. Cent., Edmonton, Alberta. Forest Manage. Note.